Macmillan/McGraw-Hill Science

FORCES & MACHINES

AUTHORS

Mary Atwater
The University of Georgia

Prentice Baptiste
University of Houston

Lucy Daniel
Rutherford County Schools

Jay Hackett
University of Northern Colorado

Richard Moyer
University of Michigan, Dearborn

Carol Takemoto
Los Angeles Unified School District

Nancy Wilson
Sacramento Unified School District

Downhill skier

Macmillan/McGraw-Hill
School Publishing Company
New York Chicago Columbus

CONSULTANTS

Assessment:
Mary Hamm
Associate Professor
Department of Elementary Education
San Francisco State University
San Francisco, CA

Cognitive Development:
Pat Guild, Ed.D.
Director, Graduate Programs in Education and Learning Styles Consultant
Antioch University
Seattle, WA

Kathi Hand, M.A.Ed.
Middle School Teacher and Learning Styles Consultant
Assumption School
Seattle, WA

Derrick R. Lavoie
Assistant Professor of Science Education
Montana State University
Bozeman, MT

Earth Science:
David G. Futch
Associate Professor of Biology
San Diego State University
San Diego, CA

Dr. Shadia Rifai Habbal
Harvard-Smithsonian Center for Astrophysics
Cambridge, MA

Tom Murphree, Ph.D.
Global Systems Studies
Monterey, CA

Suzanne O'Connell
Assistant Professor
Wesleyan University
Middletown, CT

Sidney E. White
Professor of Geology
The Ohio State University
Columbus, OH

Environmental Education:
Cheryl Charles, Ph.D.
Executive Director
Project Wild
Boulder, CO

Gifted:
Dr. James A. Curry
Associate Professor, Graduate Faculty
College of Education, University of Southern Maine
Gorham, ME

Global Education:
M. Eugene Gilliom
Professor of Social Studies and Global Education
The Ohio State University
Columbus, OH

Life Science:
Wyatt W. Anderson
Professor of Genetics
University of Georgia
Athens, GA

Orin G. Gelderloos
Professor of Biology and Professor of Environmental Studies
University of Michigan—Dearborn
Dearborn, MI

Donald C. Lisowy
Education Specialist
New York, NY

Dr. E.K. Merrill
Assistant Professor
University of Wisconsin Center—Rock County
Madison, WI

Literature:
Dr. Donna E. Norton
Texas A&M University
College Station, TX

Macmillan/McGraw-Hill School Division
10 Union Square East
New York, New York 10003
Printed in the United States of America

ISBN 0-02-274289-1 / 8

3 4 5 6 7 8 9 RRW 99 98 97 96 95 94 93

Mathematics:
Dr. Richard Lodholz
Parkway School District
St. Louis, MO

Middle School Specialist:
Daniel Rodriguez
Principal
Pomona, CA

Misconceptions:
Dr. Charles W. Anderson
Michigan State University
East Lansing, MI

Dr. Edward L. Smith
Michigan State University
East Lansing, MI

Multicultural:
Bernard L. Charles
Senior Vice President
Quality Education for Minorities Network
Washington, DC

Paul B. Janeczko
Poet
Hebron, MA

James R. Murphy
Math Teacher
La Guardia High School
New York, NY

Clifford E. Trafzer
Professor and Chair, Ethnic Studies
University of California, Riverside
Riverside, CA

Physical Science:
Gretchen M. Gillis
Geologist
Maxus Exploration Company
Dallas, TX

Henry C. McBay
Professor of Chemistry
Morehouse College and Clark Atlanta University
Atlanta, GA

Wendell H. Potter
Associate Professor of Physics
Department of Physics
University of California, Davis
Davis, CA

Claudia K. Viehland
Educational Consultant, Chemist
Sigma Chemical Company
St. Louis, MO

Reading:
Charles Temple, Ph.D.
Associate Professor of Education
Hobart and William Smith Colleges
Geneva, NY

Safety:
Janice Sutkus
Program Manager: Education
National Safety Council
Chicago, IL

Science Technology and Society (STS):
William C. Kyle, Jr.
Director, School Mathematics and Science Center
Purdue University
West Lafayette, IN

Social Studies:
Jean Craven
District Coordinator of Curriculum Development
Albuquerque Public Schools
Albuquerque, NM

Students Acquiring English:
Mario Ruiz
Pomona, CA

STUDENT ACTIVITY TESTERS
Alveria Henderson
Kate McGlumphy
Katherine Petzinger
John Wirtz
Sarah Wittenbrink
Andrew Duffy
Chris Higgins
Sean Pruitt
Joanna Huber
John Petzinger

FIELD TEST TEACHERS
Kathy Bowles
Landmark Middle School
Jacksonville, FL

Myra Dietz
#46 School
Rochester, NY

John Gridley
H.L. Harshman Junior High School #101
Indianapolis, IN

Annette Porter
Schenk Middle School
Madison, WI

Connie Boone
Fletcher Middle School
Jacksonville, FL

Theresa Smith
Bates Middle School
Annapolis, MD

Debbie Stamler
Sennett Middle School
Madison, WI

Margaret Tierney
Sennett Middle School
Madison, WI

Mel Pfeiffer
I.P.S. #94
Indianapolis, IN

CONTRIBUTING WRITERS
Elizabeth Alexander
Gene Seabolt

ACKNOWLEDGEMENTS
Reprinted with permission of Atheneum Publishers, an imprint of Macmillan Publishing Company from *THE GUARDIAN OF ISIS* by Monica Hughes. Copyright © 1981 Monica Hughes.

Reprinted with permission of Charles Scribner's Sons, an imprint of Macmillan Publishing Company from *WINDMILLS, BRIDGES AND OLD MACHINES* by David Weitzman. Copyright © 1982 David Weitzman.

A rower uses machines to move her boat.

Forces and Machines

EXPLORE

TRY THIS

Features

Departments

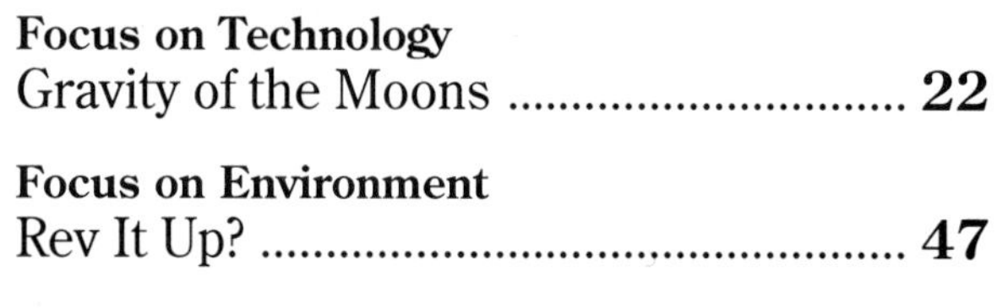

Forces AND Machines

What pictures, or images, pop into your mind when you hear the word *machine?* Perhaps you see something enormous, like a tower crane that lifts concrete slabs onto tall buildings or bridges. Perhaps you see a jackhammer chopping concrete slabs. Maybe you see something that has a gasoline engine, such as a fire engine or sports car, when you hear the word *machine.*

You may not think of yourself as a machine operator, but there are many machines you use every day. Look at the pictures on this page. They show a pencil sharpener and other items that you use so often you may not even think of them as machines.

Using a crane makes some jobs much easier to do.

A pencil sharpener does work on a pencil when you turn the crank.

Suppose the pencil sharpener in your classroom disappeared. How would you adjust? Would you use pens in every assignment, even to solve math problems? Would you take your pencils home at night and scrape away the wood with a piece of sandpaper? Imagine how much more time you would spend sharpening pencils if you had to use sandpaper. Have you ever emptied a pencil sharpener? Think about all the wood shavings it holds. What a mess sharpening pencils with sandpaper would create; the cleanup alone might double the time it takes to do the job! We seldom realize how much more difficult the simplest tasks would be without machines.

Machines don't just help us do our work. They help us when we're playing, too. Suppose you live close enough to your friend to get to her home easily by bicycle but far enough away to make walking impractical. You can cover more distance more quickly on a bicycle than on foot. Using a bicycle can give you more time for fun.

Does a bicycle move without your help? Not unless you're moving downhill. Otherwise, you have to apply a force. You push down the pedals to turn the wheels that make a bicycle move.

Think about pushes and pulls, or forces. You already know something about forces. You've studied Sir Isaac Newton's explanations for how forces affect motion. Scientists call these explanations Newton's Laws of Motion. In this unit you'll discover how the forces of friction and gravity help us move and how they get in our way. You'll find out how energy can be used to do work and how machines, such as the ones pictured on pages 6 and 7, make work seem easier to do in many different systems. You'll learn that in science "doing work" often means "having fun."

Minds On! Think about the contents of your kitchen. Are there any machines? Identify the machines in your kitchen. List the items you identify in your ***Activity Log*** on page 1. ●

Doing work on a bicycle can be fun.

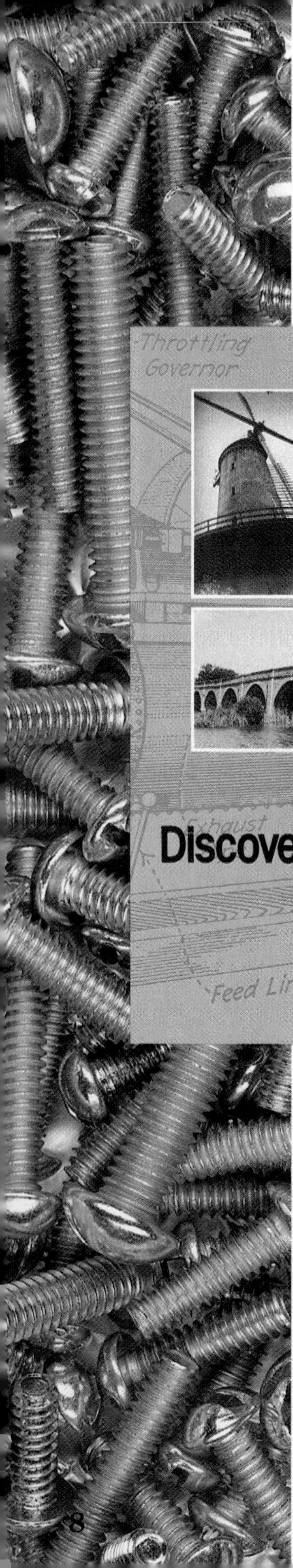

Science in Literature

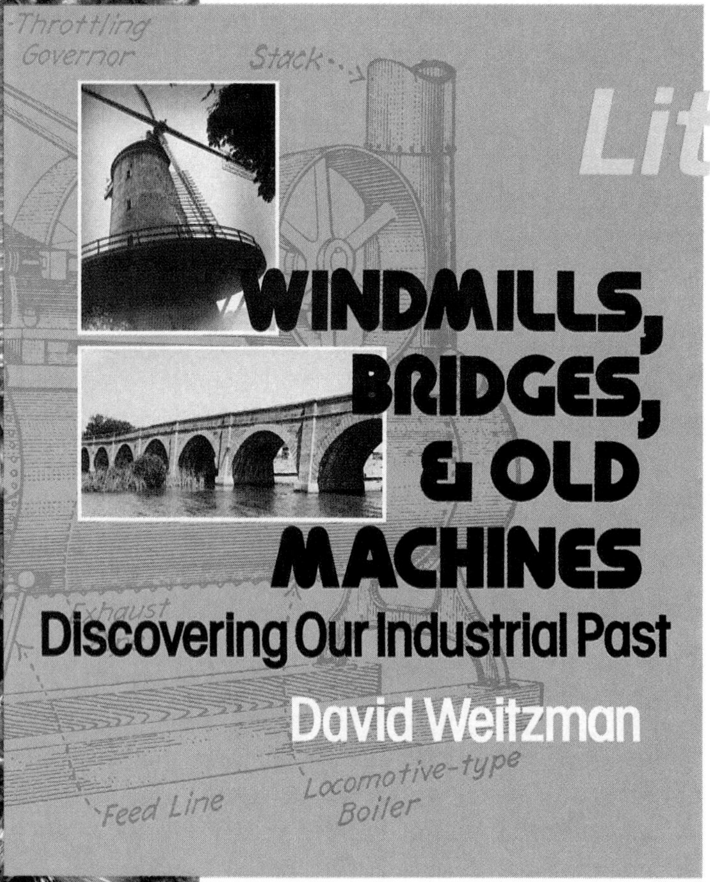

Forces and machines have been the subject of numerous works of literature. Several of these books are described for you on these two pages. You should find these books not only to be informative on the subject of forces and machines, but also to be enjoyable reading.

WINDMILLS, BRIDGES, & OLD MACHINES
Discovering Our Industrial Past
David Weitzman

Throttling Governor
Stack
Exhaust
Feed Line
Locomotive-type Boiler

Windmills, Bridges, & Old Machines: Discovering Our Industrial Past **by David Weitzman. New York: Charles Scribners Sons, 1982.**

Machines have been used since the time of the earliest humans to accomplish tasks in less time using less human energy. These tasks have included almost everything from transportation to cooking.

This book will lead you on an exploration of machines of the past. Locomotives, windmills, and foundries for making steel are all featured here. Waterwheels and many other fascinating machines have helped to make the world we live in today. Read this book to find out how machines have helped people throughout the ages accomplish everyday tasks.

The Guardian of Isis **by Monica Hughes. New York: Atheneum Press, 1981.**

Jody N'Kumo was born on Isis and had never known life on Earth. His ancestors had come to Isis from Earth, but did not want life on Isis to duplicate life on Earth. Isis did not have any simple machines, and Jody's interest in simple machines was not welcome. Read about Jody's struggles and his journey to see the Guardian.

Other Good Books To Read

The Amazing Adventures of Albert and His Flying Machine **by Thomas Sant. New York: Lodestar Books, 1990.**

Have you ever considered owning your own unidentified flying object (UFO)? What would happen if you used a UFO to deliver newspapers? Albert purchased a flying machine that many people in his hometown mistook for a UFO. You'll enjoy reading about his adventures, and you may be motivated to own a UFO yourself someday.

Pyramid **by David Macaulay. Boston: Houghton-Mifflin Company, 1975.**

More than two million blocks of stone were used to construct the pyramids that Egyptian builders built with great skill and machines. Each step in the construction of a pyramid is illustrated in clear detail in this book, making one of the world's wonders more understandable.

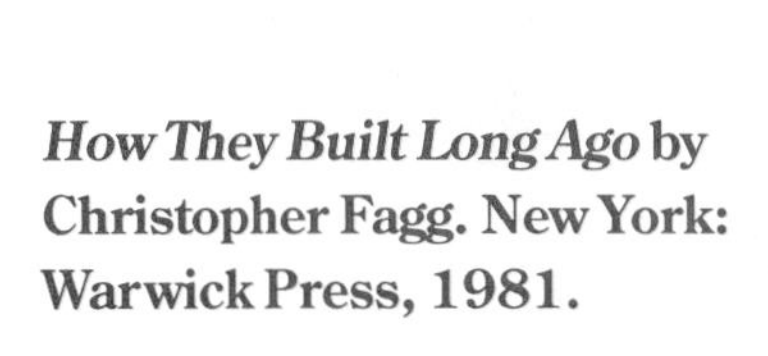

How They Built Long Ago **by Christopher Fagg. New York: Warwick Press, 1981.**

Long ago could be 4000 B.C. or A.D. 1500 and all the years in between. Many structures like Stonehenge, the Egyptian pyramids, and the Great Wall of China cause present-day engineers to marvel at their size and design. Some tools in use in ancient times bear a close resemblance to some modern tools. Nearly any famous structure from long ago is described here along with the history and knowledge of its builders.

ARE Friction & GRAVITY Always a Hindrance?

Gravity pulls us down and friction keeps us all from slipping into the ocean. Without gravity we might be like a meteor drifting through space. With gravity but without friction, we'd be constantly sliding until we all slid into the ocean.

Ronnie is a long jumper. His best jump covered more than seven meters on a dry day with very little wind. He tried several times to go more than seven meters (about 23 feet) a few days later, but the weather was stormy that day. Each time he went to hurl himself for a new personal record, one of his feet would slip on the wet surface. He tried a different pair of shoes for his final attempt. He didn't notice any slipping as he sprung into the air, but the rainy atmosphere made him feel like he had bowling balls on each shoulder. He wasn't surprised to find out that his jump barely covered six meters (about 20 feet).

Long jumpers like Ronnie depend on friction for success. When long jumpers push on Earth with their feet, they want Earth to push back on them as forcefully as possible. When the surface is slippery, this doesn't happen.

A jump's length is measured from the front edge of the takeoff board to the nearest mark made in the sand by Jackie's foot.

A competitor in the long jump runs down a long runway and leaps into a pit of sand. Jackie Joyner-Kersee is an olympic track star shown here in midair during a long jump.

Friction can also contribute to a long jumper's failure. While long jumpers are in the air, they want friction between themselves and the molecules in the air to slow them as little as possible. Increasing this air friction can shorten their jumps.

Long jumpers also have to contend with another force known as gravity. No matter how well long jumpers like Ronnie succeed in catapulting themselves skyward, gravity always brings them quickly back to Earth.

Minds On! Work with a partner. Make a list of all the things and movements you do that involve friction. Illustrate or find pictures of some of the items you mentioned. Include these illustrations and pictures with your list in your *Activity Log* on page 2. ●

EXPLORE

Activity!

Getting Things Moving

You know that sometimes there is a lot of friction. **Friction** is the force that opposes the motion between two surfaces that are touching each other. Friction between skin and carpet is great enough to cause a burn in children who go sliding across carpets on bare knees. Other times there isn't very much friction at all. Suppose a child rolls across the carpet. Would he or she be burned? Probably not. In this activity you will find out what variables affect the amount of friction between two objects. What do you think some of these variables might be? List them in your ***Activity Log.***

What You Need

2 blocks of wood (10 cm × 5 cm)
spring scale
sandpaper
2 dowel rods
10 g of baby powder
masking tape
***Activity Log* pages 3–4**

What To Do

1 Hook a spring scale to the block with sandpaper on one side, as shown in the picture. Be sure that the block is positioned with the sandpaper side facing up. Using a steady force, use the spring scale to pull the block across the table at a steady speed.

2 Record the force needed to pull the block across the table in your ***Activity Log.*** Pull the block across the table three more times. Then, calculate and record the average force needed to pull the block across the table.

3 Find out whether the weight of the block affects the amount of friction. Place the second block on top of the first one, doubling the weight. Repeat step 2 using two blocks instead of one.

4 Find out whether the type of surface affects the amount of friction. Unstack the blocks and set the second block aside. Then, turn the first block of wood over so that the sandpaper faces down. What do you think will happen to the amount of friction now? Repeat step 2 using the block with the sandpaper side against the table this time.

5 Check to see if surface area affects friction. Put the block on its side as shown in the picture. (The weight of the block is the same, but the area in contact with the table is much less—only about half as much.) Repeat step 2, using the block on its narrow side instead of its wide side.

6 You can see the effect of a lubricant on friction by sprinkling a small amount of baby powder on the table. Repeat step 2.

7 So far, you've experimented only with sliding friction. Find out whether rolling produces more or less friction than sliding by using the dowel rods. Place the block on top of the rods. Repeat step 2.

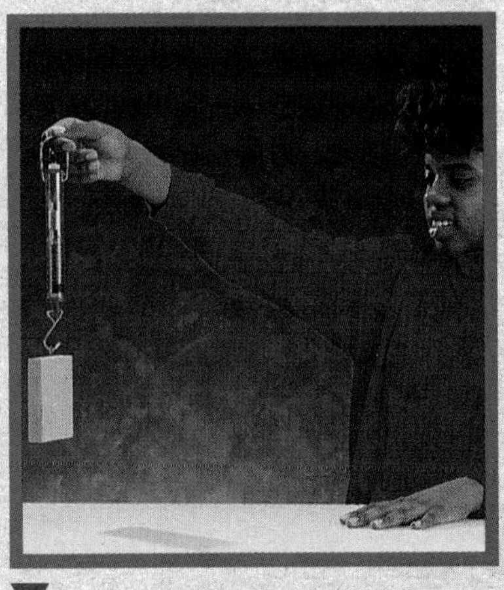

8 Find out how much your block weighs by holding it in the air with the spring scale. Record the weight in N in your ***Activity Log.***

What Happened?

1. What force are you actually measuring?
2. Which variables seemed to affect the frictional force between the block and the table?
3. Under what condition was the frictional force the smallest?

What Now?

1. In the activity, what would happen to the friction if a third block were added?
2. Why do you suppose a water slide is more slippery than a regular playground slide? Explain how you could prove your answer using the materials in this activity.
3. Ramps with rollers are frequently used to unload boxes from a truck. What advantage does a ramp with rollers have over a ramp without rollers?

A Course About Force

Even though friction may appear to be a force that always gets in the way, we often need it. Without it, cars, trucks, and bicycles would merely spin their wheels. Houses and most kinds of furniture wouldn't hold together because the nails would slide out of the boards. Without friction you couldn't hold paper in your fingers or cassette tapes in your hands. Any attempt at squeezing a baseball would result in it shooting from your fingers! Even walking wouldn't be possible. Without friction you would just slip and slide. What are some other activities that you need friction to do?

Like humans, animals are dependent upon friction, too. Have you ever seen a snake strike at its prey? Perhaps you've noticed that it rears up before it strikes. But, in order for it to do this, there must be friction. On slippery surfaces, such as a sheet of ice, a snake can't strike. Friction's opposing force provides the support a snake needs. Without it, the snake couldn't get the food it needs to survive.

Snakes like this rattlesnake might starve if it weren't for friction.

Friction opposes any attempt you make to move the surface of one object over another. In the Explore Activity, you moved the surface of the block over the surface of the table and measured different amounts of opposing frictional force. You saw that the force of friction is affected by the type of material (sandpaper versus wood) and the force between the two objects. In the Explore Activity, the force between the objects was the weight of the block.

When you moved the block with the sandpaper side up, the frictional force was less than with the sandpaper side down. The type of material affects the force of friction.

When you turned the block on its side and pulled it across the table, you discovered that surface area does not affect friction.

Friction is needed for holding nails in boards, keeping trucks on roads, and for just about everything else we do. Sometimes we want to reduce friction. In the Explore Activity, you saw one method that people have used for thousands of years to reduce friction—rolling objects instead of dragging them. Look at the data in the table in your ***Activity Log.*** Your investigations with the wooden blocks, table, and spring scale revealed that rolling produces less frictional force than sliding does. Think about how much we rely on wheels to reduce the forces required to move things. Wheels allow us to roll rather than slide. Make a list of some things that use wheels.

What happened in step 6 of the Explore Activity? When you sprinkled baby powder on the table before pulling the block across it, you used lubrication to reduce the friction. As you may have experienced, a water slide is more slippery than a regular playground slide for the same reason. The water lubricates the slide, or decreases the amount of friction that may exist between you and the slide. Oil also reduces friction between surfaces. Perhaps you have used oil on a door hinge or on your bike chain. People have used lubrication for a long time. More than 4,000 years ago, the ancient Egyptians lubricated large blocks of stone with oil so that the blocks could be moved to build statues and pyramids.

The ancient Egyptians knew that oil could be used to reduce friction.

Language Arts Link

The Real McCoy

You know that friction can be useful as well as harmful. At one time Elijah McCoy had worked for the railroad, shoveling coal and oiling the trains' moving parts. He decided that a device could be invented to automatically oil the parts of a machine. He began working on the problem. In July of 1872, Elijah McCoy received a patent for the first automatic lubricator, named the "lubricator cup." The device let small amounts of oil drip onto the moving parts of a machine. In later years, anyone who had a McCoy lubricator bragged that it was the "real McCoy," an expression still used to mean something of quality.

Write a one- to two-page fiction story about what life might have been like if Elijah McCoy hadn't been born.

Elijah J. McCoy—1843–1929

More force is needed to pull the block when the friction is greater.

Look again at the data table you made in the Explore Activity. Remember that each time you pulled the block across the table, you read a measurement from the spring scale. The measurements ranged from 0 to 20 newtons. What do these numbers mean?

Think about the friction you encountered when you pulled the block across the table. In step 1 you felt little friction as you pulled a single wooden block across the table. In step 4, when you turned the block so that the sandpaper side faced down, you felt more friction. In every step, you pulled the block in one direction while friction pulled against you in the opposite direction. The numbers you read from the spring scale showed the amount of force you used to overcome friction and move the block.

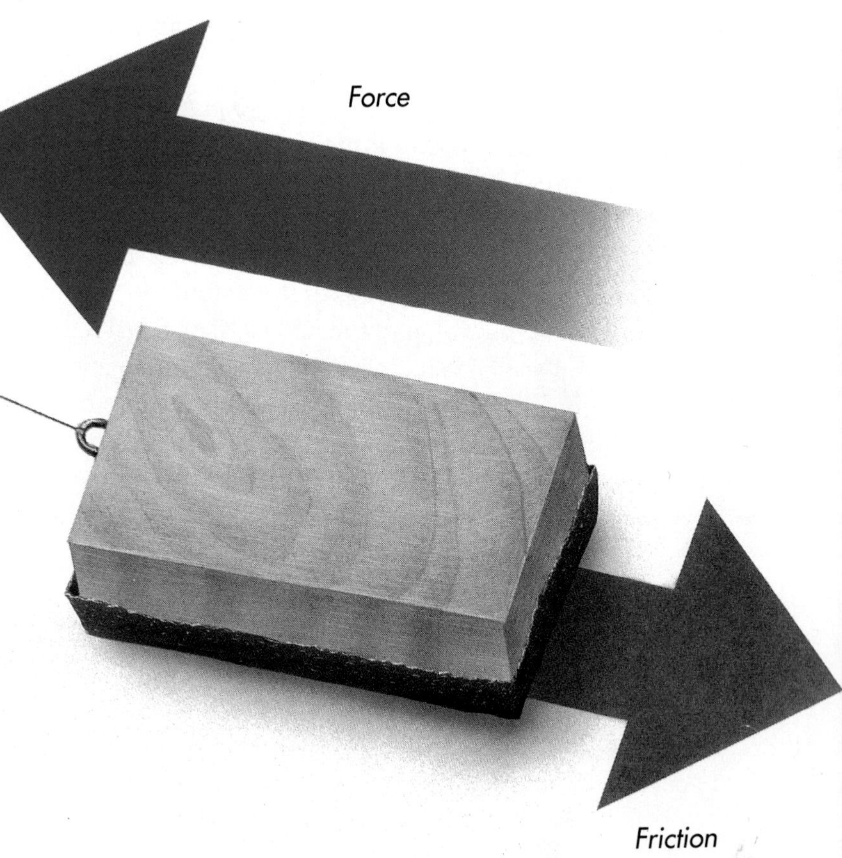

Suppose that when you pulled the block with the sandpaper side up, you obtained a measurement of three newtons. That means it took three newtons of force to move the block at a steady speed. How much force did it take to move the stacked blocks at a steady speed? Find the answer in your data table.

Have you ever played tug-of-war? Think about the rope in a tug-of-war game. Suppose the forces exerted by the two teams on the rope are the same. Will either team win? If the force that one team applies to the rope equals the force the other team applies, then a state of balance, or equilibrium, exists. The rope won't move.

Newton's first law of motion tells us about what happens to an object when the forces acting on it are balanced. Whatever the object was doing, it just keeps on doing the same thing. So if it was stationary, or not moving, it stays stationary; it won't start to move. But if it was already moving, it keeps moving in exactly the same direction in a straight line as long as the forces acting on it are balanced.

Minds On! Think about sitting on your bicycle at a stoplight. You know that you won't start moving unless you or someone else exerts a force to start you moving. The forces acting on you and the bicycle have to become unbalanced for you to start moving. This is probably pretty obvious to you. But now think about what happens as you are pedaling along at a constant speed. Do you think of the forces acting on you and the bike as being balanced? What are these forces? Describe these forces in a paragraph. ●

You probably listed **gravity,** or the attraction between Earth and an object, as pulling you toward Earth. What keeps you from falling into the center of Earth? You don't fall into the center of Earth because the pavement pushes upward with just enough force to balance the downward force of gravity. You probably described the force caused by your pedaling. Did you list friction? What happens if you stop pedaling? You gradually coast to a stop because the frictional forces acting on you and the bike from the air and road are no longer balanced by your pedaling. So while you're going at a constant speed, the force you cause by your pedaling just balances the frictional forces, and Newton's first law of motion is evident. You keep going in a straight line at the same speed until an unbalanced or net force acts on you.

Water in this stream forms a waterfall because of gravitational force.

If the force of gravity acting on this sky diver weren't being reduced by the air resistance acting on the parachute, impact with Earth's surface could cause serious injury or death.

Newton's second law of motion has to do with what happens when the forces acting on an object are not balanced. When there is a net force, an object changes its motion. An object can change its motion in many ways. It can speed up, slow down, or stop, and it can change the direction in which it is moving. How fast it does any of these things depends on two factors. The first is how large the net force is. The bigger this force, the greater the change in the motion of an object. The other factor is the amount of matter, or mass, of the object. A net force causes less change of motion for a greater mass than a smaller mass.

Think about pedaling your bike again. When do you pedal the hardest, to start and get up to speed, or when you're moving along at constant speed? When are the forces acting on your bike balanced, and when are they not balanced? What does the unbalanced force of your pedaling when you start cause you and your bike to do? Now think about what happens when you want to stop quickly. What do you do?

Minds On! Describe how the forces are balanced or unbalanced as you come to a stop on your bike. What if you were carrying a large, heavy pack on your back and started up on your bike? Describe how this would feel different. Would you be able to get up speed as quickly? How is this an example of the effect of mass in Newton's second law? Write a paragraph that includes your descriptions and explanations. ●

In the Explore Activity, you investigated how friction opposes motion. You know that when you walk, climb, stretch, or jump, you struggle against gravity as well. Gravity always pulls objects in the same direction—toward the center of Earth. People and other organisms, oceans, air, and everything else in the world around us are held in place by gravity.

Isaac Newton discovered that the force of gravity between two objects depends on two things—mass and distance. Weight is the measure of the force of gravity on an object. The force of gravity on Earth's surface is larger than the force of gravity on the moon's surface. Which has more mass, Earth or the moon? Where would your weight be greater, on the surface of Earth or the surface of the moon? The answers to these questions give you a clear picture of how the force of gravity increases when mass increases.

You can understand how distance affects the force of gravity by comparing the weight of the same elephant in San Diego, California, and in Denver, Colorado. Denver is farther from the center of Earth than San Diego is. If an elephant with a mass of 5,000 kilograms (about 11,000 pounds) were transported from the San Diego Zoo to the Denver Zoo, the weight of the elephant would change. In San Diego the elephant would weigh 49,000 newtons, but in Denver the same elephant would weigh only 48,950 newtons. The weight difference of 50 newtons is not due to the elephant's refusing to eat any hay while in transport. Everything, including elephants, weighs less in Denver than in San Diego.

If an elephant moves farther from the center of Earth, it will lose weight.

Maybe it seems odd that gravity pulls on an elephant with less force when the elephant is farther from the center of Earth. Do the Try This Activity below to observe another odd situation that involves gravity.

TRY THIS Activity!

Rolling Uphill

You expect gravity to cause objects to go downward. In this activity you'll see something unexpected.

What You Need

board, 2 nickels, cardboard cylinder, masking tape, *Activity Log* page 5

Do a demonstration that seems to defy gravity. Prepare a gently sloped ramp using a board and books. Tape 2 nickels inside a short cardboard cylinder cut from a tube used to hold roll flooring. Place the cylinder on the ramp so that the mass is on the uphill side. Let go of the cylinder. What happens? How can you explain what you saw? Write an answer in your ***Activity Log.***

Newton's first law of motion tells us that when an object is moving at constant speed in a straight line, all of the forces acting on the object are balanced. In the Explore Activity, when you pulled the block forward, you pulled with just enough force to balance the force of friction that was pulling backward on the block. When you changed the friction between the block and the table by using sandpaper, baby powder, or dowels you noticed a change in the reading on the spring scale. You had to pull less if friction were less, and more if friction were greater. If there were no friction between the block and the table would you read any force on the spring scale? Why?

Look at the data on ***Activity Log*** page 3. How did the force required to pull the block across the table compare with the weight of the block? What force do you balance when you lift the block? In which direction does the opposing force pull? When you pulled on the string in the Explore Activity you exerted a force. According to Newton's third law, the block exerted a force on you in opposite direction. Remember that forces always act in pairs.

Throughout time people have looked for ways to overcome the forces of gravity and friction, or for ways to use these forces to help them with daily tasks. Read the Literature Link to find out more about these inventions.

Literature Link

Windmills, Bridges, & Old Machines

In the 1800s before railroads and paved roads had been constructed, travel was difficult. You may have read of pioneers and covered wagons or open carts drawn by horses. Try to imagine how it would be to drag a cart full of heavy furniture across a marshy landscape.

One way to decrease friction between a mass and Earth was to move the mass by boat. But how would you get up and down hills in a boat? Canals and locks provided the answer.

Locks are easily understood if you think of them as a flight of water stairs going up a hill or down, if you're going that way (they work both ways). Instead of having the water and canal boats run downhill, the engineer puts in a lock or flight of locks that "step" the boat along with the water down the slope. (page 24)

Read the book *Windmills, Bridges, & Old Machines* by David Weitzman. When you have finished the book, choose one of the inventions in the book and describe it in detail. How does it use friction or gravity to accomplish its goal? How does it use other forces to counteract gravity? What does it use to reduce or enhance friction? When you are finished, design your own device to accomplish a task such as lifting a weight from the ground that uses or overcomes the forces of friction and gravity.

Using a Force To Stay on Course

Tires on bicycles, automobiles, and trucks have treads to increase friction. Suppose you are riding a bicycle with smooth tires and your friend has a trail bike with rough tires. If both of you raced across a muddy football field, who would be likely to win? You might not get past the starting line because the lack of friction between your tires and the field could leave you spinning your wheels. On the other hand, your friend's tires would dig into the mud, and the force of friction would push the bike forward.

Tire treads funnel away water to maintain friction between the surfaces of tires and roads in rainy weather. With treads, your bicycle tires *can* grip road surfaces in the rain. You can stop and start the bike as needed. You have more control.

The treads on tires can give a vehicle more friction and more control. Read the information on the next page to learn ways that scientists are planning to use gravity to control how a space vehicle travels.

Which tire would give more friction?

Focus on Technology

Gravity of the Moons

The spacecraft *Galileo* is on its way to Jupiter. In 1995 it will explore Jupiter's moons. Scientists who have planned the mission of the spacecraft *Galileo* are using the force of gravity among the four moons of Jupiter to give the craft the energy it will need to "dance" among them. Engineers have plotted the mission carefully so that *Galileo* will pass close by one moon at just the right moment for the gravitational force of that moon to pass it on to the next moon or "dance partner." Perhaps you're wondering how gravity can work as a force to move spacecrafts. If you remember, every object exerts a gravitational force on every other object. The strength of the force between *Galileo* and the moons of Jupiter will be strong enough to boost the spacecraft from moon to moon around the planet.

If you read the science sections of a newspaper or magazine, you may see the results of *Galileo's* mission. The spacecraft will take close-up pictures of Jupiter and its moons. We may discover things about the outer planets that neither you, nor astronomers, nor other scientists ever imagined—thanks to gravity.

Moons of Jupiter

Gravitational force of moon applied to Galileo spacecraft
Galileo

The gravity of a moon will pull on the Galileo as it approaches the moon.

Sum It Up

In this lesson you have learned how the forces of friction and gravity can help and hinder movement in different systems. You know two methods by which sliding friction can be reduced—reducing the weight and lubricating the surface. You've seen how gravity holds objects on Earth and how it can be used to boost objects in deep space.

Laws of motion first stated by Isaac Newton describe how forces such as friction and gravity can change the motion of an object. How much change in motion occurs and how long motion continues to change depends on something besides force. In Lesson 2 you will learn how energy affects motion.

Using Vocabulary

friction
gravity

Draw a picture or write a sentence that correctly explains the vocabulary words.

Critical Thinking

1. Tonia and Elena use the same sidewalk when they walk to school. One morning after a freezing rain, the sidewalk was covered with ice. Elena was wearing rubber-soled gym shoes, and Tonia had on shoes with smooth leather soles. As they were walking, Tonia complained that she could not walk without sliding. Elena didn't think it was that slippery. Use what you have learned about friction to explain what each student was experiencing.

2. Why would you weigh more on Mt. Everest than you would at a low elevation on the moon?

3. Why is it more difficult to roller skate up a hill than it is to walk up the same hill?

4. From what you have learned about friction, why do trains have wheels instead of lubricated tracks for moving loads?

5. A car in neutral and parked on a slope will roll down the slope. If the parking brake is engaged, the car doesn't move. Explain this situation.

HOW Does Energy CHANGE?

If you have a lot of energy, do you find it easy to laugh? When your trainload of energy is gone, maybe you sleep or cry. Does energy come and go? Actually, energy just changes. In this lesson you will look at energy and how it changes.

Snow skiers take advantage of the small frictional force between snow and metal. Have you ever tried snow skiing? If you have, you've experienced the rapid acceleration that can happen when gravity pushes you with little opposition from friction. Unless you learn how to make more friction by snowplowing or zigzagging, you'll soon find yourself traveling so fast that you might lose control.

At the bottom of a ski slope, you usually find a long, level surface. If this surface is long enough, you'll come to a stop without plowing into the snow or falling down. If this surface begins to take you uphill, you could find yourself stopping and then sliding backward.

Minds On! Name some activities other than snow skiing where you use gravity to rapidly accelerate with little opposition from friction. Also, describe how you are able to stop after gravity stops accelerating you. Include these activities and descriptions in your ***Activity Log*** on page 6.

Was roller skating among the activities you mentioned? Do the Explore Activity to study forces that act on a roller skate.

Gravity can accelerate a skier to an uncomfortable speed.

EXPLORE

Activity!

The Energy for Starting and Stopping a Ride

How does the force of friction opposing the motion of an object compare with the force of gravity that started it in motion? Do mass and height affect the motion of a skate rolling down a ramp?

What You Need

roller skate
1-m × 20-cm board
meterstick
3 wooden blocks
scissors
spring scale
strapping tape
***Activity Log* pages 7–8**

What To Do

1 Find the force of friction opposing the motion of a roller skate by hooking a spring scale to a roller skate and pulling the skate at a steady speed. Record the reading on the spring scale in your ***Activity Log.*** Now, hold the roller skate in the air with the spring scale to find its weight. Record this reading also.

2 Make a ramp for accelerating a roller skate by placing a wooden block under one end of a board. Use the meterstick to measure the height from the floor to the upper surface of the board on the raised end. Record this measurement in your ***Activity Log.***

3 Place the roller skate on the ramp so that the rear wheel is at the raised end of the ramp. Let the skate roll down the ramp and across the floor. Be sure the skate does not hit any obstacles. Measure the distance from the bottom of the ramp to the rear wheel of the skate on the floor at the place where it stopped rolling. Write this measurement in your ***Activity Log.***

4 Now, fasten a wooden block to the roller skate using a piece of strapping tape. Repeat step 1 to find the force of friction and the weight of the skate-block combination.

5 Repeat step 3 to find how far the skate-block combination rolls across the floor.

6 Make the ramp higher by adding another block under the board. Measure the height of this higher ramp as you did in step 2.

7 Repeat step 3 using the higher ramp and the skate-block combination.

What Happened?

1. Using the lower ramp, which rolled farther across the floor, the roller skate or the skate-block combination?
2. Which ramp caused the skate-block combination to roll farther across the floor, the lower ramp or the higher ramp?

What Now?

1. Where did the roller skate receive the energy needed for it to roll across the floor?
2. Why did the roller skate stop rolling?
3. What factors influence how far the skate will roll across the floor?

Two Kinds of Energy

In the Explore Activity, a force acted over a certain distance to make the roller skate move and to cause it to stop. In each case work was done on the roller skate. **Work** is defined as a force being applied through a distance. When you placed the roller skate at the top of the ramp, you had to apply a force over a distance. To find out how much work you did, you would need to use a mathematical equation.

$$WORK = FORCE \times DISTANCE$$

You would measure force in newtons and distance in meters. Multiplying newtons times meters would give a quantity called a newton·meter. Another name for a newton·meter is a joule. The **joule** is the metric unit of work or energy and is equal to a force of one newton applied over a distance of one meter.

Exactly how much work is one joule? Imagine that you lift a stick of butter from one shelf to another that is one meter higher. Have you done any work? How much? It might not seem like very much work, but you did need to exert a force of about one newton over a distance of one meter. The 1-newton force was needed to overcome the force of gravity pulling down on the stick of butter. So, how much work have you done?

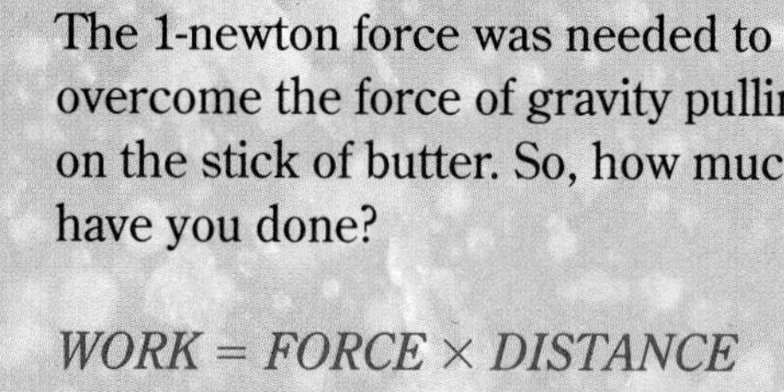

$$WORK = FORCE \times DISTANCE$$

$$W = F \times d$$

$$W = 1 \text{ newton} \times 1 \text{ meter}$$

$$W = 1 \text{ newton·meter (or 1 joule)}$$

How much work would you do if you lifted two sticks of butter one meter?

$$W = F \times d$$

$$W = 2\text{ N} \times 1\text{ m}$$

$$W = 2 \text{ joules}$$

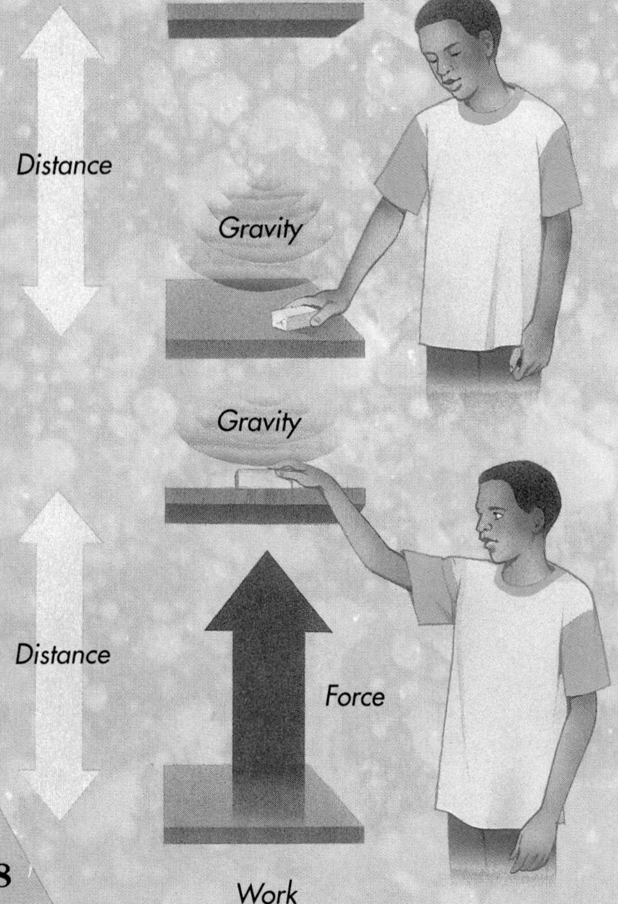

A stick of butter weighs about one newton. Lifting a stick of butter for one meter would be about one joule of work.

Where did you do work on the roller skate in the Explore Activity? How much work did you do? When you placed the roller skate on the top of the ramp, you had to exert a force equal to the weight of the roller skate. You worked against the force of gravity for a distance equal to the height of the ramp. Suppose the roller skate weighed 3 newtons and the ramp had a height of 0.15 meters.

$$W = F \times d$$
$$W = 3\text{ N} \times 0.15\text{ m}$$
$$W = 0.45\text{ joule}$$

Perhaps you are thinking, 3 newtons is larger than 1 newton. Why did it take more work to lift the stick of butter? Remember, work is a product of force and distance. The stick of butter was lifted for a much greater distance.

When you lift an object such as a roller skate or a stick of butter to a higher position, you give a type of energy called potential energy to the object. **Potential energy** is the stored energy that an object has because of its position or condition. It didn't require much energy or work to lift the roller skate to the top of the ramp. The roller skate at the top of the ramp does not have very much potential energy. If you were to lift a heavy rock to the top of a stepladder, the rock would have much more potential energy than the roller skate at the top of the ramp.

You can think of many additional examples of potential energy. A skier at the top of a mountain has potential energy, as does a cliff diver leaning over the edge of a cliff and a rock poised to fall down the side of a hill. Water held by a dam in a reservoir also has potential energy. This water will flow to a city that is located below the reservoir without using energy from a pump.

These rocks are found in Arches National Park in Utah. Think about the potential energy they have.

Minds On! How does the condition of an object affect its potential energy? Think about the shape of a spring like the one that is inside the spring snakes pictured on these pages. Suppose that the spring has been compressed inside the can. What will happen when the can is opened? Because of the condition of the spring when it is compressed, it has potential energy. If the can is opened, the spring will act to restore itself to its original condition. Can you think of any objects that use the stored or potential energy of a spring? Try to find pictures of these objects. Record these objects with the pictures you could find in your ***Activity Log*** on page 9. ●

You've already learned that work is the product of force times distance ($W = F \times d$). When you calculated the work done to lift a roller skate or a stick of butter, you used the weight of the object as the force and the height to which it was lifted as the distance. Your work gave gravitational potential energy to the object. The amount of gravitational potential energy of an object is a direct function of the object's weight and its height. When did the roller skate have the energy to roll farther? It rolled farther when you increased its weight or lifted it higher. Gravitational potential energy is the product of weight times height.

$$\text{Gravitational Potential Energy} = \text{Weight} \times \text{Height}$$
$$PE = Wt \times h$$

In the Explore Activity, you gave gravitational potential energy to the roller skate when you lifted it onto the ramp. Notice that this is only one form of potential energy. You may recall the different forms of energy, such as mechanical energy, chemical energy, electrical energy, thermal energy, and radiant energy. Each of these forms of energy can be stored as potential energy. What form of energy is stored when a spring is stretched? When food is eaten?

Will the energy stored in these spring snakes make them jump?

If potential energy is the type of energy that is stored, there must be another type of energy. Think back to the Explore Activity where you stored energy in the roller skate by lifting it onto a ramp. What happened to the roller skate as a result of this stored energy? It rolled or went into motion. The stored energy was changed into energy of motion. Another name for the energy of motion is **kinetic energy.** Which roller skate would have more kinetic energy, the rolling skate at the bottom of the lower ramp, or the rolling skate at the bottom of the higher ramp? Since the skate was given more potential energy when it was lifted to the higher ramp, the rolling skate should have more kinetic energy at the bottom of the higher ramp. Did raising the ramp make the roller skate look like it was moving faster when it reached the bottom? When an object is moving faster, its velocity has increased. You may recall that velocity is a measure of an object's speed and direction. Increasing velocity increases kinetic energy. Do the Try This Activity on the next page to observe two ways to increase kinetic energy.

Jumping spring snakes have energy of motion.

TRY THIS Activity!

Box Bowling

You know that a moving object has energy of motion. What effect does a change in mass or velocity have on the kinetic energy of a moving object?

What You Need

empty aluminum soft drink can, full aluminum soft drink can, board, 3 wooden blocks, empty 16-oz cornstarch box or similar box, *Activity Log* page 10

Place one block under the board. Put the empty box at the low end of the board on the floor. Try to knock the box over using the full soft drink can rolling down the ramp. Now, try knocking the box over using an empty can of the same volume. Add blocks under the ramp one at a time until the empty can moves fast enough to knock over the box. Write your observations in your ***Activity Log.***

In the Try This Activity above, you observed how two variables, mass and velocity, affect the energy that an object has and the strength of the force that it can exert. The full soft drink can has much more mass than the empty one. The velocity of the cans describes how fast they travel in the direction of the box. The more mass or velocity a can had, the stronger the force that it could exert to upend the box.

You may have been surprised that the empty can was able to upend the box. Increasing the velocity of an object has a greater effect on its kinetic energy than increasing its mass does. Doubling the velocity of the can actually quadruples its kinetic energy. Tripling the velocity would increase the kinetic energy by nine times. The relationship among kinetic energy, mass, and velocity can be expressed mathematically—kinetic energy = one-half the mass times the square of the velocity.

$$KE = \frac{1}{2} \times m \times v^2$$

A skateboarder knows how energy changes feel.

Can a skateboarder know how far he or she will roll?

Where Are the Missing Joules?

Think about the energy changes or conversions that occur when you bounce a rubber ball. When you hold the ball in your hand, before you let go, what type of energy does it have? You release the ball. It falls to the ground. As it is falling, what two types of energy does it have? When it strikes the floor, what type of energy does it have?

The ball bounces back up after it strikes the floor. As it bounces back up, the energy is again converted. When the ball reaches the top of its bounce, does it have any energy?

When you are holding the ball in your hand before you let it go, it has potential energy. As it is falling, it has both potential and kinetic energy. Since it can still fall some more, it has potential energy, and its velocity gives it kinetic energy. When it strikes the floor, its energy is entirely kinetic, and its velocity reaches its largest value. As it bounces back up, its velocity gradually decreases. This decreases its kinetic energy, but the rise in height increases its potential energy. At the top of the ball's bounce, its energy is entirely potential.

You know from experience that the ball will not bounce back to the point from which it was dropped. Why? Some of the ball's energy was "lost" due to the opposition of friction as it moved through the air. Some kinetic energy was converted to thermal energy and sound when the ball struck the floor.

A bouncing ball experiences energy changes.

If you were to add the energy that was converted to heat and sound and that worked against friction, it would account for the difference in height after the bounce. The total amount of energy remains the same. This concept is known as the law of conservation of energy. Because the amount of energy is the same before and after a change, we say energy has been conserved. Do the following Math Link to find out if energy was conserved when you rolled the skate.

Braking a Skate

What happened to the potential energy of the roller skate in the Explore Activity? Did the work of the skate against friction as it rolled to a stop across the floor equal the potential energy that the skate had before it began to roll? Find your measurements in your ***Activity Log.*** Add your measurements to the sample measurements.

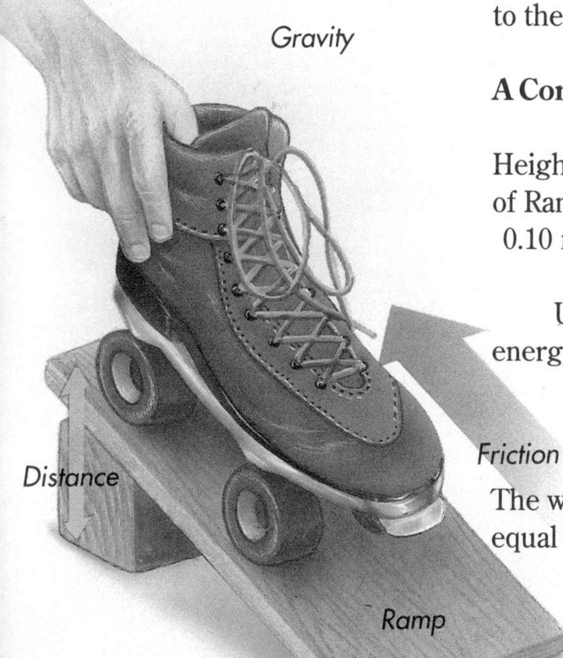

This skate has potential energy at the top of the ramp. It changes to kinetic energy as it moves down the ramp.

A Comparison of Potential Energy and Work Against Friction

Height of Ramp	Weight of Skate	P.E.	Force	Distance	Work of Friction
0.10 m	3 N		0.5 N	0.55 m	

Using the sample data from the table above, potential energy at the top of the ramp would equal

$$PE = Wt \times h$$
$$PE = 3\text{ N} \times 0.10\text{ m}$$
$$PE = 0.30\text{ joule}$$

The work of the skate against friction as it rolls to a stop would equal

$$W = F \times d$$
$$W = 0.5\text{ N} \times 0.55\text{ m}$$
$$W = 0.28\text{ joule}$$

To compare the potential energy and the work against friction, subtract the work from the P.E.

$$0.30\text{ joule} - 0.28\text{ joule} = 0.02\text{ joule}$$

Now, do the same calculations using the measurements you made during the Explore Activity. Was energy conserved? How do you account for any energy that seems to have been lost?

Sum It Up

In this lesson you constructed a skate-ramp system and experimented with it to find how energy changes as a skate rolls to a stop. You learned that the weight of an object and its height affect gravitational potential energy. You saw how potential energy can transform to kinetic energy. When potential energy changes to kinetic energy and back to potential energy, as it does when you bounce a ball, energy is conserved.

Using Vocabulary

joule **potential energy**
kinetic energy **work**

Write a definition for energy using each of these words.

Critical Thinking

1. Aaron can push a car on a level surface. If he pushes the car for a distance of 10 meters from one level place to another, has he increased the potential energy of the car? Why or why not?
2. When a basketball is going through the net, what kind of energy does it have?
3. Carlita can climb a rope from the gym floor and touch the ceiling, which is 10 meters above the gym floor. If Carlita is 2.5 meters tall and she weighs 450 newtons, how much work does she do when she climbs the rope?
4. If you stretch a bow with an arrow, release the arrow, and the arrow hits the bull's-eye on a target, when does the arrow have the maximum potential energy? The maximum kinetic energy?
5. If a roller coaster rolls down a steep place in the track, why won't it climb to a height that is higher after reaching the bottom of the steep place?

Does Work Gain Momentum If the Power Increases?

Have you ever attempted to ring a bell that hangs high above the ground by tossing a stone at the bell? The stone does work on the bell when it strikes it, causing the bell to move. You would have to do enough work on the stone to give it enough momentum after it leaves your hand to reach and ring the bell.

In everyday life people use the word *work* in a variety of ways. They often talk about having a lot of work to do (or a lot of homework), or they say they are getting ready to go to work. People may also say that they are tired because they worked so hard. In a scientific sense, these situations may or may not be examples of work. In Lesson 2 you learned that in science, work is defined as a force being applied through a distance. If you lift a stick of margarine one meter (about three feet), you are doing work as science defines the term.

A chisel does work when a wood-carver uses a force to move it.

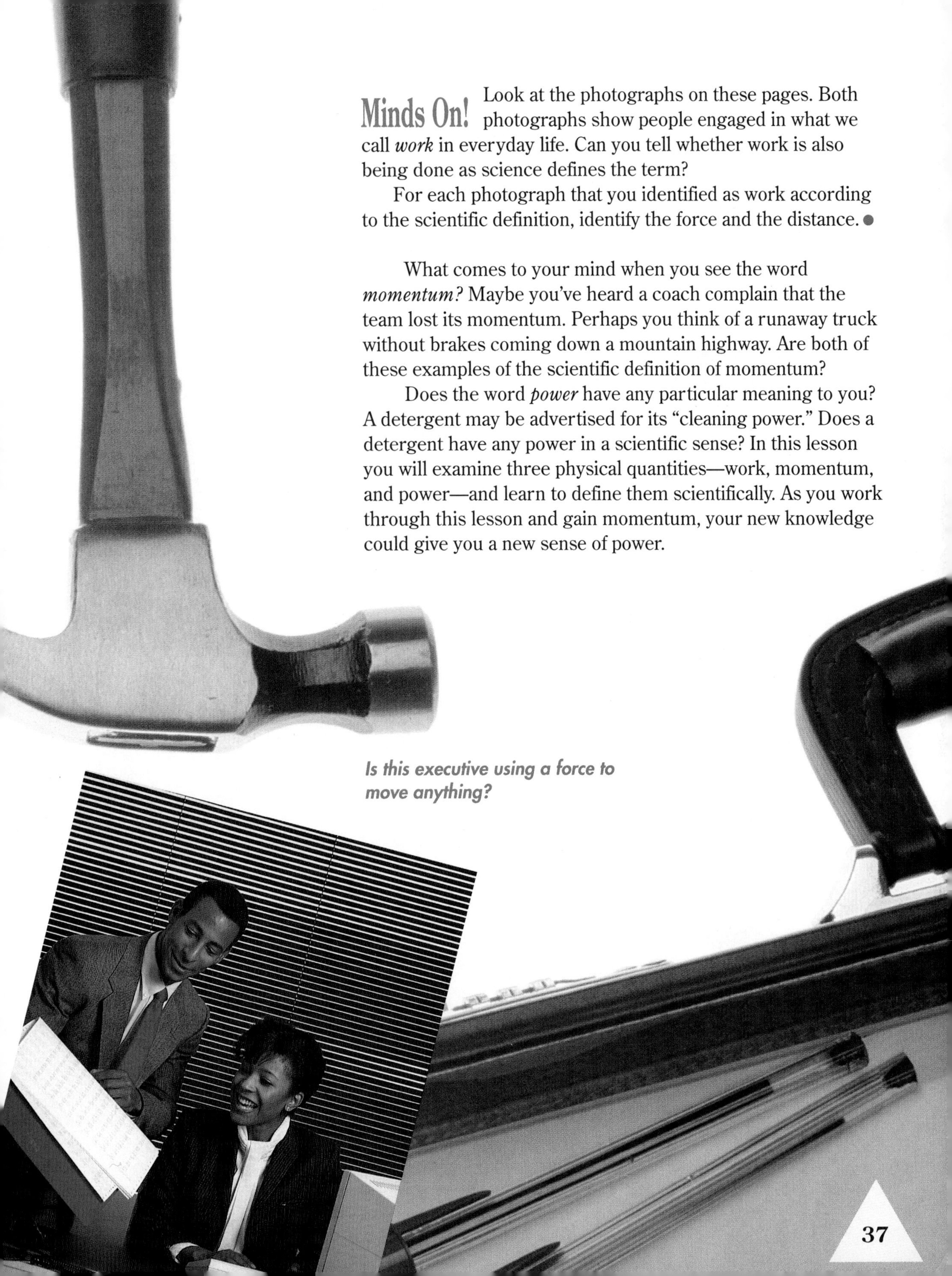

Minds On! Look at the photographs on these pages. Both photographs show people engaged in what we call *work* in everyday life. Can you tell whether work is also being done as science defines the term?

For each photograph that you identified as work according to the scientific definition, identify the force and the distance. ●

What comes to your mind when you see the word *momentum?* Maybe you've heard a coach complain that the team lost its momentum. Perhaps you think of a runaway truck without brakes coming down a mountain highway. Are both of these examples of the scientific definition of momentum?

Does the word *power* have any particular meaning to you? A detergent may be advertised for its "cleaning power." Does a detergent have any power in a scientific sense? In this lesson you will examine three physical quantities—work, momentum, and power—and learn to define them scientifically. As you work through this lesson and gain momentum, your new knowledge could give you a new sense of power.

Is this executive using a force to move anything?

EXPLORE

Activity!

Lift Yourself Up

Do you remember that in Lesson 2 you calculated the amount of work done on an object? Now, you're going to find out the rate at which an object does work. You won't be experimenting with just any old object—the object is you!

What You Need

meterstick
stopwatch
stairs
***Activity Log* pages 11–12**

What To Do

1 Find a staircase where you won't bother any classes.

2 You need to measure the vertical height of the stairs. This can be done by measuring the height of one step as shown in the picture and then multiplying by the total number of stairs. Record the vertical height of the stairs in your ***Activity Log.***

3 Now you need to find your weight, the force that you must exert to lift yourself, in newtons. Either use a scale that measures weight in newtons or multiply your weight in pounds by 4.5. Record your weight in newtons in your ***Activity Log.***

4 Ask a friend to use the stopwatch to time you as you walk up the stairs. Repeat to find an average time.

5 Calculate how much work you did. Remember that work equals force times distance.

6 Calculate the rate at which you worked by dividing the amount of work you did by how many seconds it took.

What Happened?

1. How much work did it take to climb the stairs?
2. At what rate did you walk up the stairs?

What Now?

1. Look back at page 28 and think about lifting a stick of butter 1 m. What would your rate of work be if it took you 1 s to lift the butter? If it took 2 s?
2. Try to think of a way to measure the rate of work of a small, wind-up toy car moving *up* a ramp. Record your answer in your ***Activity Log.***

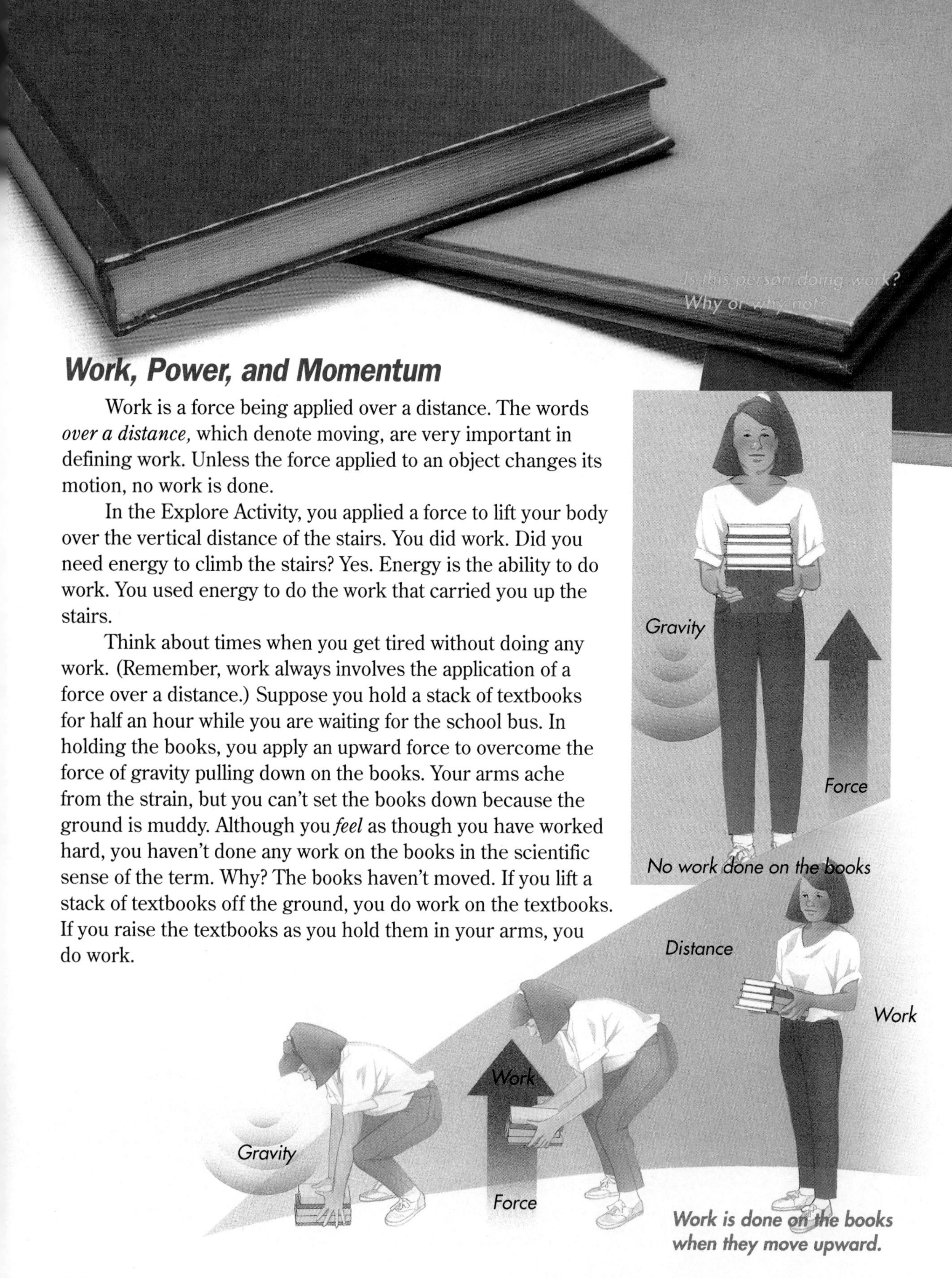

Work is done on the books when they move upward.

Work, Power, and Momentum

Work is a force being applied over a distance. The words *over a distance,* which denote moving, are very important in defining work. Unless the force applied to an object changes its motion, no work is done.

In the Explore Activity, you applied a force to lift your body over the vertical distance of the stairs. You did work. Did you need energy to climb the stairs? Yes. Energy is the ability to do work. You used energy to do the work that carried you up the stairs.

Think about times when you get tired without doing any work. (Remember, work always involves the application of a force over a distance.) Suppose you hold a stack of textbooks for half an hour while you are waiting for the school bus. In holding the books, you apply an upward force to overcome the force of gravity pulling down on the books. Your arms ache from the strain, but you can't set the books down because the ground is muddy. Although you *feel* as though you have worked hard, you haven't done any work on the books in the scientific sense of the term. Why? The books haven't moved. If you lift a stack of textbooks off the ground, you do work on the textbooks. If you raise the textbooks as you hold them in your arms, you do work.

Remember how in the Explore Activity you calculated your rate of work? You divided the amount of work that you did climbing stairs by the amount of time required for you to make the climb. Refer to the time measurements in your ***Activity Log*** on pages 11 and 12 that show how fast you worked as you climbed the stairs. Compare the rates of work that you recorded with those of some classmates. Notice that some rates of work are slightly higher than others.

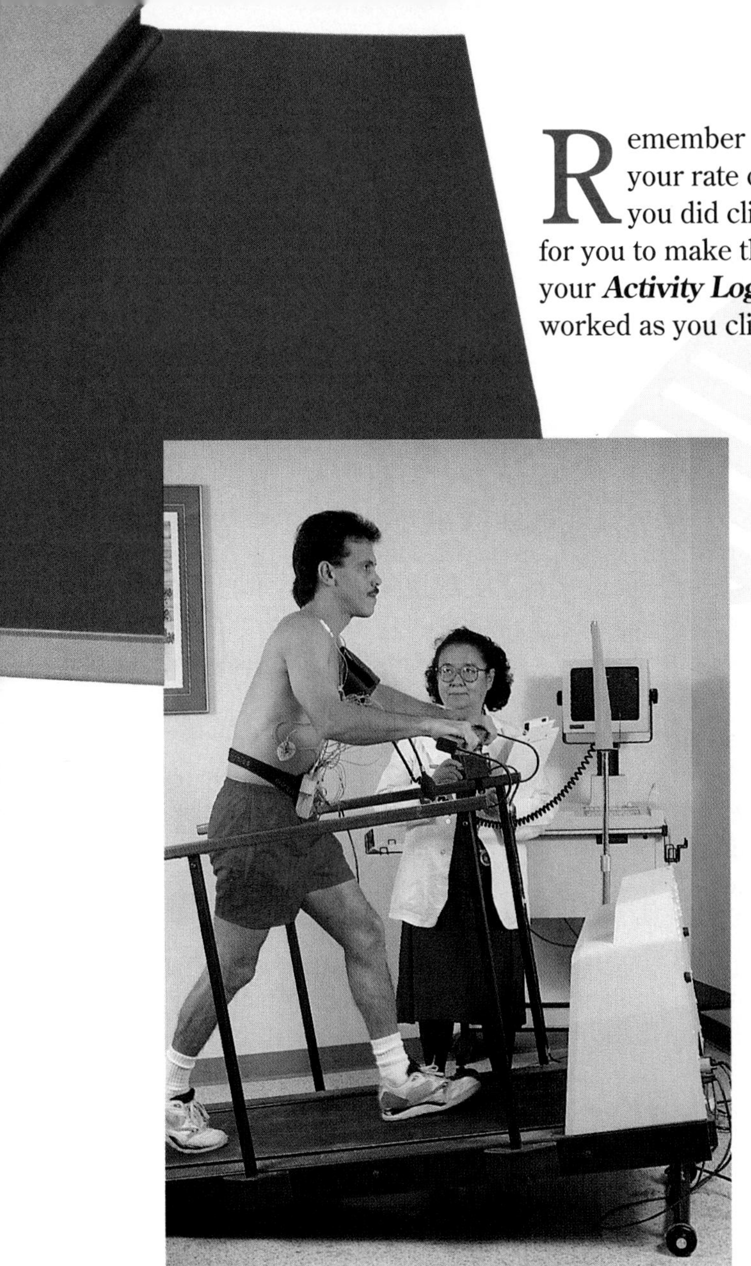

This machine can measure a person's power.

In science **power** is defined as the rate at which work is done. You were calculating power when you divided your amount of work by the amount of time it took you to climb the stairs. You know that when you do work, you use energy. Power can also be defined in science as the rate at which energy is used. Power is measured in units called watts. A **watt** is equal to one joule of work per second (J/s). Look at the data you recorded in your ***Activity Log*** during the Explore Activity. How many watts of power did you use when you walked up the stairs?

Minds On! Perhaps you have heard your parents or other adults use the word *horsepower* in reference to car engines. One horsepower is equal to approximately 746 watts.

Originally, horsepower referred quite literally to how fast a horse could do work. The British inventor James Watt, for whom the *watt* is named, coined the term *horsepower.*

Refer once again to the data you recorded in the Explore Activity. Calculate how much horsepower you generated when you walked up the stairs. Now calculate how fast you would have had to climb the stairs to generate one horsepower. Include this calculation in your ***Activity Log*** on page 13. ●

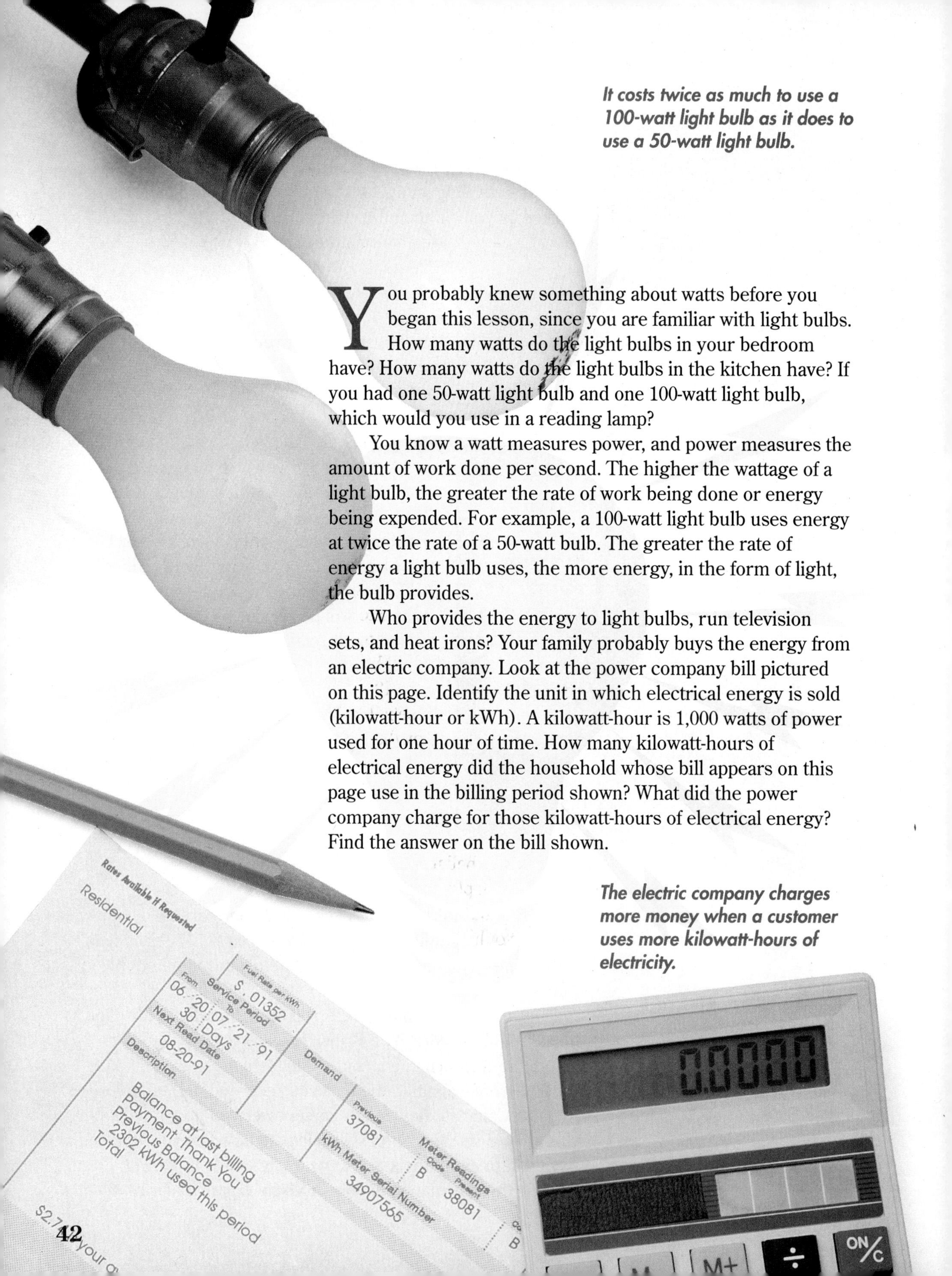

It costs twice as much to use a 100-watt light bulb as it does to use a 50-watt light bulb.

You probably knew something about watts before you began this lesson, since you are familiar with light bulbs. How many watts do the light bulbs in your bedroom have? How many watts do the light bulbs in the kitchen have? If you had one 50-watt light bulb and one 100-watt light bulb, which would you use in a reading lamp?

You know a watt measures power, and power measures the amount of work done per second. The higher the wattage of a light bulb, the greater the rate of work being done or energy being expended. For example, a 100-watt light bulb uses energy at twice the rate of a 50-watt bulb. The greater the rate of energy a light bulb uses, the more energy, in the form of light, the bulb provides.

Who provides the energy to light bulbs, run television sets, and heat irons? Your family probably buys the energy from an electric company. Look at the power company bill pictured on this page. Identify the unit in which electrical energy is sold (kilowatt-hour or kWh). A kilowatt-hour is 1,000 watts of power used for one hour of time. How many kilowatt-hours of electrical energy did the household whose bill appears on this page use in the billing period shown? What did the power company charge for those kilowatt-hours of electrical energy? Find the answer on the bill shown.

The electric company charges more money when a customer uses more kilowatt-hours of electricity.

In baseball if a batter hits with power, does the batter do a large amount of work in a short period of time? Does a power pitcher have a rapid rate of doing work? Both the batter and the pitcher do work on the baseball when they accelerate the baseball. To figure out how much work is done you must calculate the force applied to the baseball. Force equals the mass of an object times its acceleration. For example, a pitcher might be able to accelerate a baseball from 0 meters per second to 40 meters per second in one second. If the mass of the baseball is about 0.2 kilograms, the force applied to the baseball by the pitcher would be found by using the formula.

$$F = m \times a$$
$$F = 0.2\ \text{kg} \times 40\ \text{m/s/s}$$
$$F = 8\ \text{N}$$

The pitcher is only applying a force to the baseball from the place where the ball begins accelerating to the place where the pitcher releases the ball. If this distance is about two meters, the work done on the baseball by the pitcher would equal 16 joules.

$$W = F \times d$$
$$W = 8\ \text{N} \times 2\ \text{m}$$
$$W = 16\ \text{joules}$$

This work is done in about one second, so the pitcher's power is 16 watts.

$$P = W/t$$
$$P = 16\ \text{J}/1\ \text{s}$$
$$P = 16\ \text{watts}$$

You probably had more power when you climbed the stairs. Are you a power climber?

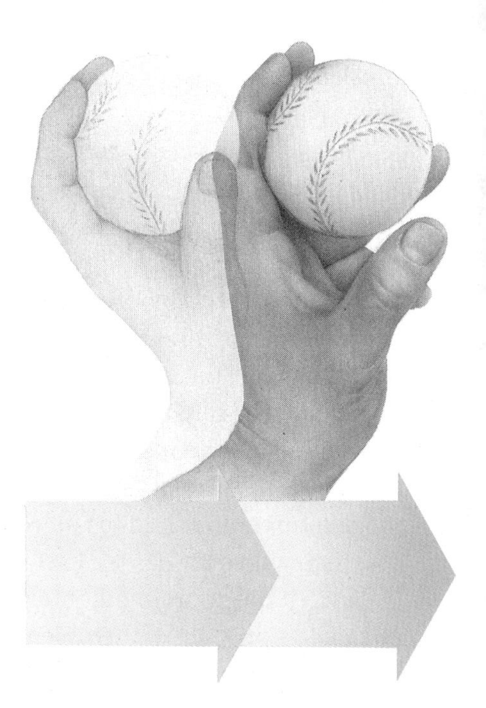

A pitcher applies force to a ball until it is released. After that point, it continues in the direction of the applied force until the forces of gravity and friction stop it or it is hit or caught.

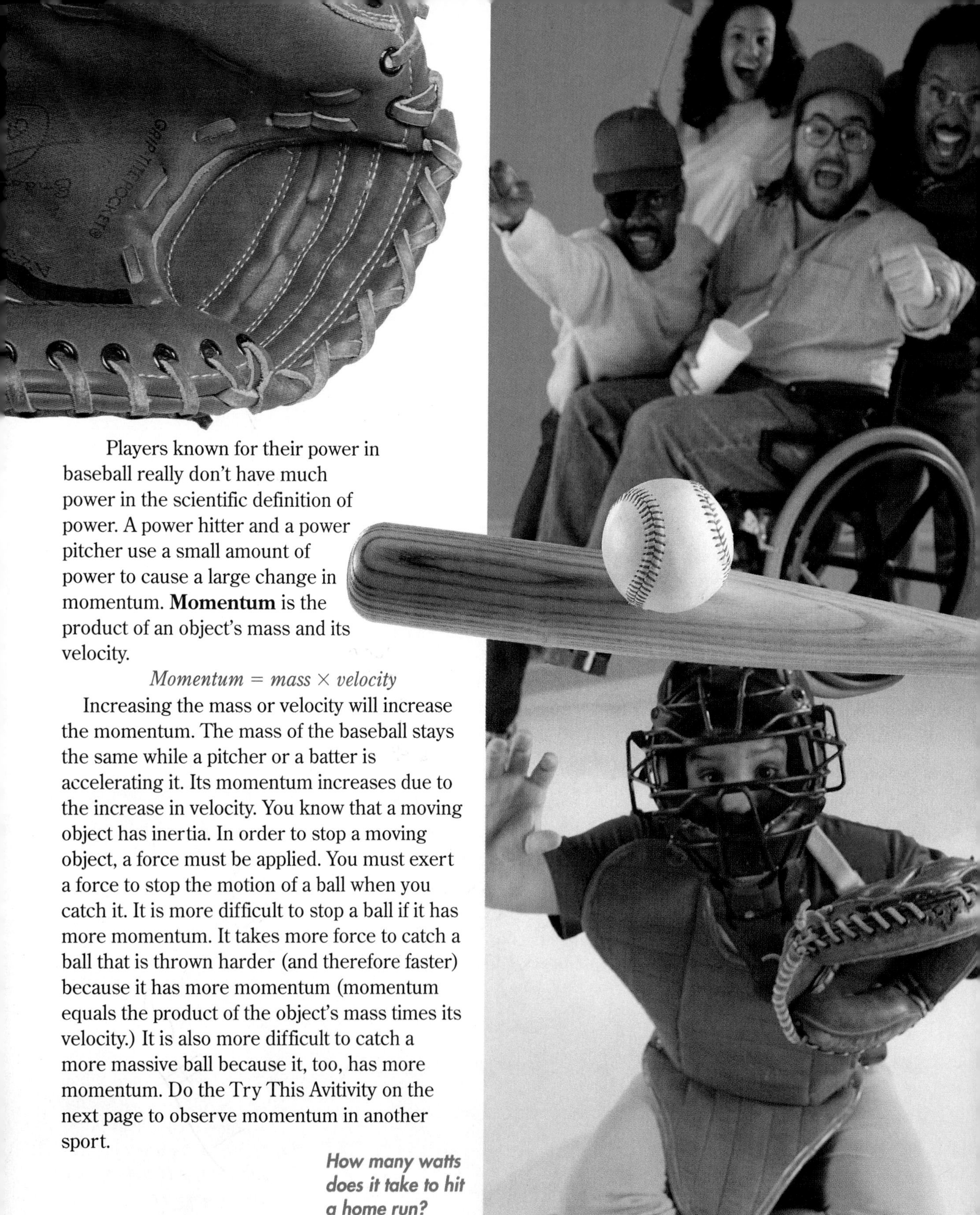

Players known for their power in baseball really don't have much power in the scientific definition of power. A power hitter and a power pitcher use a small amount of power to cause a large change in momentum. **Momentum** is the product of an object's mass and its velocity.

$$Momentum = mass \times velocity$$

Increasing the mass or velocity will increase the momentum. The mass of the baseball stays the same while a pitcher or a batter is accelerating it. Its momentum increases due to the increase in velocity. You know that a moving object has inertia. In order to stop a moving object, a force must be applied. You must exert a force to stop the motion of a ball when you catch it. It is more difficult to stop a ball if it has more momentum. It takes more force to catch a ball that is thrown harder (and therefore faster) because it has more momentum (momentum equals the product of the object's mass times its velocity.) It is also more difficult to catch a more massive ball because it, too, has more momentum. Do the Try This Avitivity on the next page to observe momentum in another sport.

How many watts does it take to hit a home run?

Let It Roll

How does the velocity of a basketball as it is dropped affect its momentum?

What You Need

basketball
basketball hoop
gym shoes
stopwatch
***Activity Log* page 14**

Have a friend with a stopwatch stand under one basket on a basketball court. Obtain a basketball and stand under the opposite basket holding the ball. When your partner says "go," walk toward your partner carrying the ball with you. As you cross the mid-court line, drop the ball (don't pass it). Your partner should start the stopwatch when you drop the ball and stop the watch when the ball crosses the boundary line behind the basket. Record the time in your ***Activity Log.***

Repeat this procedure once again, but this time *run* to the mid-court line before dropping the ball. Make sure you drop the ball before you begin to slow down. Record the time for the ball to cross the boundary line in your ***Activity Log.*** Did the ball cross the boundary line with more momentum this time? Explain your answer using the definition of *momentum.*

According to Newton's first law, a basketball will continue in motion if it is dropped while you are moving. If it were not for the force of friction from the floor and the air opposing the basketball's forward motion, it would never stop. Since the basketball had a relatively small mass and it wasn't traveling at a very high velocity, you probably were not concerned about the basketball knocking a hole in a wall. Sometimes, objects can have very large momentums, and considerable force is required to bring the objects to a stop. Read the Health Link on the next page to learn one way that damage and injury from objects with large momentums is avoided.

Unless a basketball is moving, it has no momentum.

Bumpier Roads and Smoother Vehicles

Health Link

Speed Bumps

Have you ever been riding in a car after dark and suddenly felt a jolt that nearly sent you through the roof? (Fortunately, you were wearing a seat belt.) You then realized that the road had "bumps" in it that were put there deliberately. The driver of your car did not see the bumps until it was too late. Do "speed bumps" have any real value?

Speed bumps are usually found on roads where people, especially young children, frequently need to cross the road. A car or truck can't travel more than 45 kilometers per hour (about 28 miles per hour) on a road with speed bumps without the passengers experiencing a high degree of discomfort. By discouraging vehicles from traveling at speeds in excess of 45 kilometers per hour, the speed bumps make it much easier for vehicles to come to a complete stop. Safety experts have compiled statistical charts that clearly show that a car or a truck is much more likely to stop in time if the vehicle is traveling at a speed below 45 kilometers per hour.

Bumps that decrease a car's velocity also decrease its momentum.

Obtain a statistical chart that gives you the average stopping distances required for cars and trucks traveling at different speeds. Some possible places to obtain one of these charts are local traffic police departments, state highway patrol depots, and driving schools. Make a diagram of a football or soccer field on a piece of paper. Draw lines on your diagram corresponding to the stopping distances on the chart. Think of some places where you have seen or felt speed bumps. Suppose there were no speed bumps on these roads. If you were riding in a car and saw a small child crossing the street only 20 meters (66 feet) in front of the car, what is the maximum speed that the car could be traveling in order to stop in time to avoid hitting the child? Include your answer and diagram in your ***Activity Log*** on page 15.

Focus on Environment

Rev It Up?

You know energy is required to do work. Where does a car obtain the energy it needs to oppose the frictional forces on roads and carry passengers like you over a distance?

Perhaps you know the term *fossil fuels.* Coal, oil, and natural gas are called fossil fuels because they are literally millions of years old, having been formed from the remains of plants and animals that lived in prehistoric times. You learned in the last lesson that the amount of energy in a system remains the same, although its form can change.

Approximately 90 percent of the energy we use in the United States comes from fossil fuels. Burning gasoline to drive cars makes up a huge part of our fossil-fuel consumption. The problem is that fossil fuels are not renewable. Consequently, scientists and engineers are striving hard to develop renewable forms of energy that *can* be replaced, and to design machines such as cars that require less use of nonrenewable energy sources.

The photographs on this page show popular cars in the United States in 1955 and today. Notice that the modern cars are much more streamlined. Why? Do you think the more streamlined design has any effect on the amount of fuel the car uses?

Working in groups of three or four, design the fuel-efficient, streamlined car that you and others may be driving in the year 2020. Use the photographs on this page as a source of ideas. Describe your ideal car in a paragraph or drawing, explaining ways its design will make the car more fuel efficient.

Down feathers, contour feathers

Are you a bird watcher? Have you ever observed the body of a bird? If so, perhaps you have observed that birds are more streamlined in shape than the most fuel-efficient car pictured on page 47. Look at the bird shown on this page. Notice its streamlined body shape. Think about how a bird moves through the air. The shape of its body minimizes the amount of friction that it creates as it moves.

A bird has two kinds of feathers—contour feathers and down feathers. Down feathers keep a bird warm. The contour feathers on a bird's tail and wings help it fly by providing lift and balance. They give a bird its streamlined shape and decrease friction produced by the air as it flies.

This lightweight aircraft was piloted nonstop around the world.

In 1983 when he was 18 years old, Jonathan Santos won a prize for a fuel-saving wingtip attachment that he had invented for airplanes. Normally air flows around the wingtip of a plane in such a way as to produce a drag. Drag makes planes less fuel-efficient. The device that Jonathan invented changes some of the force producing the drag into a force that pushes and lifts. With Jonathan's wingtip attachment, the amount of fuel needed to fly a plane could be decreased by up to 27 percent.

Jonathan began his investigations into aerodynamics when he was 14 years old. "I wanted to take flying lessons," Jonathan explained, "but they were too expensive. [I thought that] if I could invent something that would make flying cheaper, maybe I could afford the lessons."

Jonathan began by observing how birds fly. Next, he studied how air flows over aircraft wings. Finally, Jonathan applied what he had learned to design various flying gadgets and test them in a homemade wind tunnel.

Jonathan Santos observed birds in flight while designing his wingtip attachment.

A wheelbarrow can be used to move heavy loads more conveniently.

Sum It Up

In this lesson you have increased your understanding of work, momentum, and power. You observed the relationship between work, force, and distance as you climbed stairs, measured the distance over which you did that work, and calculated your rate of work, or power. Although you probably didn't get tired in the Explore Activity, you saw that energy is required to do work. When you calculated your power, you determined the rate at which your energy was used. Power can give momentum to an object. Sometimes the object can have so much momentum that it is difficult to bring it to a stop.

You know that you can use energy from your muscles to do work. Suppose you wanted to move a 100-kilogram (about 220-pound) load of wood from a truck to a fireplace in your home. Lifting the wood would take a greater force than your muscles could exert. You could move the load of wood, however, by using a wheelbarrow. In the next two lessons, you'll learn how wheelbarrows and other simple machines help people do work.

Using Vocabulary

momentum
power
watt

Describe a race between a gasoline-powered car made in the 1950s and a modern electric car using each of these words.

Critical Thinking

1. Gonzalo climbed a ladder to an open window in order to enter a room on the second floor of his house. He thought that if he used the ladder, it would require less work to get to this room than it would if he used the stairs and had to walk a longer distance. How is Gonzalo mistaken?
2. An elevator weighs 2,000 newtons and it can lift ten people who average 800 newtons apiece. If the motor on the elevator can lift this load to a height of ten meters in ten seconds, how much power does the motor have?
3. How many horsepower does the motor for the elevator in question 2 have?
4. Yoko and Marisa both love to play softball. One day they were playing catch, and Marisa could catch Yoko's fastest pitch with little difficulty. Yoko borrowed a shotput from the track team and threw it slower than she had been throwing the softball. However, Marisa fell over when she tried to catch and hold on to the shotput. Explain Marisa's different experiences with the softball and the shotput.
5. When an electric company bills a residence for the number of kilowatt-hours of electricity used, is this a bill for electric power or electric energy? Explain.

How Do Levers HELP People Do Work?

Rowers use machines to move this boat.

Could you knock on someone's door without using a lever? Many doors furnish you with a metal lever to announce your presence. These levers do the job, but you may choose to just use the lever you brought with you. Your arm can do the job just as well, maybe even better.

The pictures on this page show some of the many machines that make our lives seem easier. Try to imagine what life would be like without any machines. Could you pry the top off a bottle with your fingers? Could you dig a nail out of a board with your fingers? The force you can exert is not great enough to pry the top off a bottle. Your fingers don't have the necessary leverage to remove a nail from a board. Machines make it easier for us to do these and many other jobs.

In order for you to do any work, what must you do? Remember, the scientific definition for work is a force being applied over a distance. If you tried to pull a nail from a board with your fingers, the force of friction between the nail and the board would oppose the pull from your fingers. Unless you succeeded in pulling the nail for a measurable distance, you would not do any work on the nail, although you may feel as if you have worked very hard. A machine like a crowbar or a hammer would make it possible for you to apply sufficient force to move the nail for the distance you want to move it.

Nuts crack more easily when you use a machine.

Minds On! Look at the machines on this page. What work would you want to do using each machine? Describe the force and the distance for the work that you would use each machine to do in your ***Activity Log*** on page 16.

In this lesson we will look at several machines that transfer force from you to an object that you want to move for a measurable distance. Your understanding of how work is done will help you understand how these machines make work seem easier.

Some bottles are difficult to open without using a machine.

EXPLORE

Activity!

Hoist It Up!

You already know it is often helpful to use a machine to help you exert a force. In this activity you will explore how a lever changes the amount of force you must use.

What You Need

spring scale
meterstick
ring stand
small piece of string
large washers
large paper clip
***Activity Log* pages 17–18**

What To Do

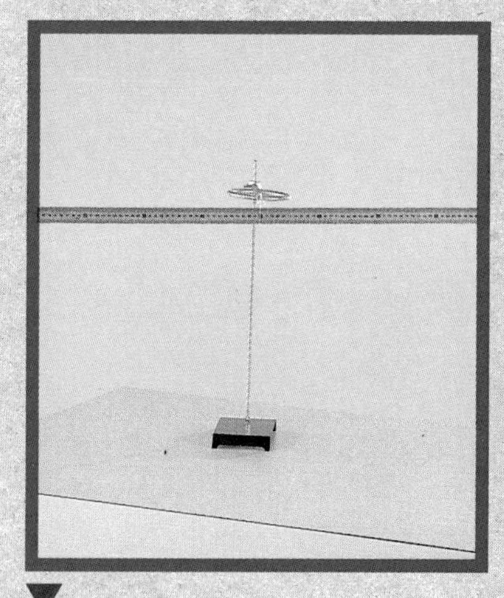

1 Use the string to tie the meterstick to the ring stand so it is balanced as shown.

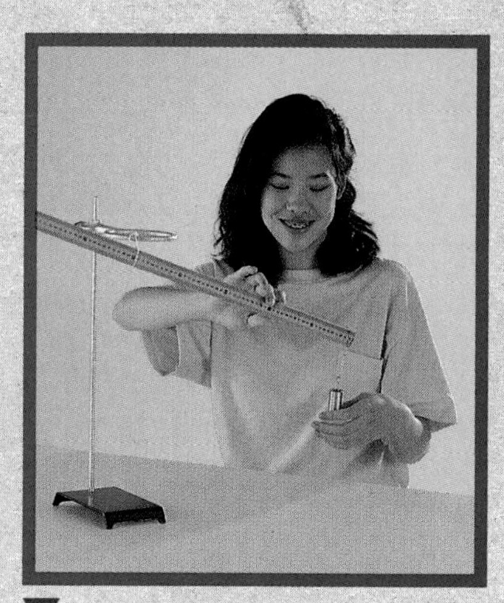

2 Use the paper clip to hang the washers from the meterstick, about 40 cm from the string.

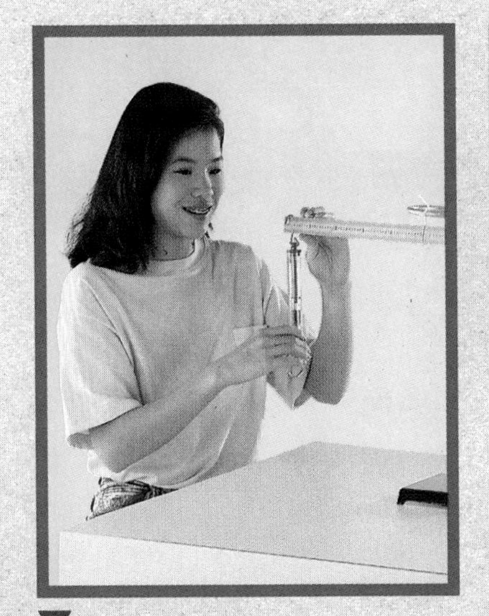

Distance from string to weights (cm)					
Weight of the washers (N)					
Distance from string to scale (cm)					
Scale reading (N)					
How far washers moved (cm)					
How far scale moved (cm)					

3 Attach the spring scale 40 cm from the pivotal point at the other end of the meterstick and pull down until the stick is level. How much force did you have to use to balance the washers? Use the table in your ***Activity Log*** to record your data.

4 Repeat step 3, placing the washers at 25 cm from the string and the spring scale at 40 cm, 30 cm, 20 cm, and 10 cm from the string until you have completed the data table. Record your results.

5 Measure how far the washers move from the at rest position in each case. Measure how far you have to pull the spring scale in each case. Record your results.

What Happened?

1. How did the lever change the force needed to lift the washers?
2. How did this particular lever change the direction of your force compared to the direction of the opposing force?
3. Did you find any time when the force you applied was just about equal to the opposing force of the washers? When?
4. What did you observe about the distance the scale moved in each case compared to the force you had to exert?

What Now?

1. Why would you ever want a lever that increased the amount of force needed to overcome the opposing force?
2. Give an example of using a lever where the distances from the opposing force to the pivotal point and the lifting force to the pivotal point are the same.

Levers can be used as seesaws.

Getting a Handle on Levers

In the Explore Activity, you used a ring stand, weights, and meterstick to make a **lever,** a bar that pivots about a fixed point. You used the lever to exert a force. Devices, such as levers that you can use to exert a force, are called **simple machines.** There are two main classes of simple machines—levers and inclined planes. You will study inclined planes in Lesson 5.

The **fulcrum,** referred to as the pivotal point during the Explore Activity, is the fixed point of your lever. The force resisting the person using the lever is called the **resistance force** (the weights in the Explore Activity). The distance from the resistance force to the fulcrum is the **resistance arm.** The person or object using the lever pushes or pulls with a force known as the **effort force.** The distance from the effort force to the fulcrum is the **effort arm.**

Look at the seesaw pictured on the next page. The seesaw is also a lever. The fulcrum is the place where the seesaw rests on the pole. Suppose one of the people on the seesaw is you. The resistance force is the weight of the person you are trying to lift. The effort force is your weight. Assume you and the other person on the seesaw weigh the same amount.

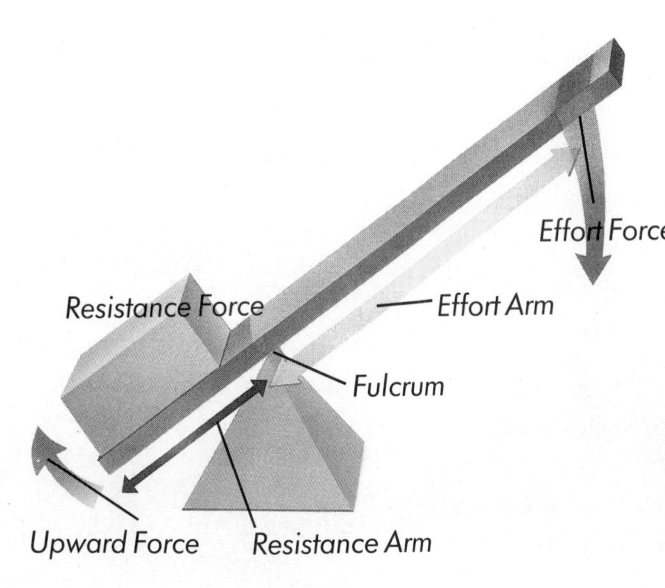

Now, think about the direction of the force you exerted on the effort arm in the Explore Activity. What happened to the resistance arm when you pushed down on the effort arm? When you pushed down, the resistance arm moved upward. The lever changed the direction of the force you applied. The seesaw works the same way. When you push up from the ground, the person on the other end of the seesaw goes downward. The seesaw changes the direction of the applied force.

Remember what happened in the Explore Activity when you moved the resistance (weights) closer to the fulcrum? The level of effort force you exerted to balance the lever decreased. When you moved the effort (meterstick and scale) closer to the fulcrum, an opposite adjustment was required. You had to increase the effort force you exerted to balance the lever.

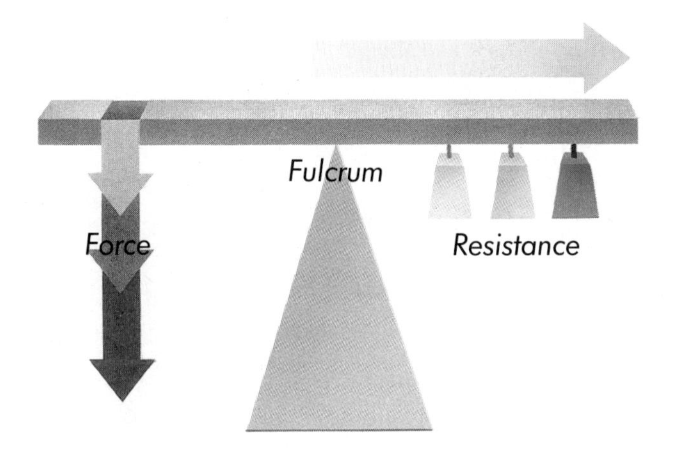

Force must increase if the resistance moves farther from the fulcrum. It decreases if the resistance moves closer to the fulcrum.

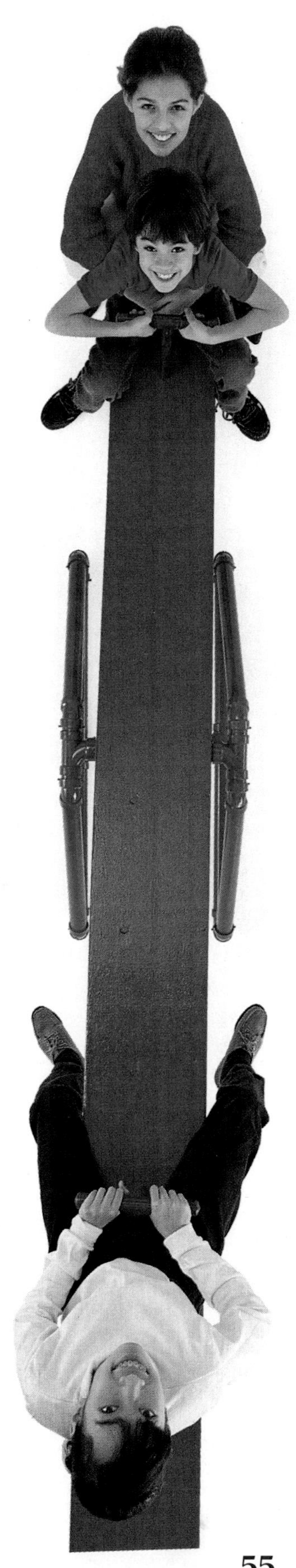

Which way would she need to move after her partner's little brother climbed onto the seesaw?

Minds On! Suppose your seesawing partner's little brother climbed onto the resistance arm of the seesaw in the example above. In what direction would you need to scoot to rebalance the seesaw? ●

Another way to understand how the distance and force exerted by a lever are related is to see how work is performed by a lever. As you learned in Lesson 2, in science work is defined as the product of force and distance. You also know that energy is conserved. This means you can't get any more energy, or work, out of a lever than you put in. Work is simply the transfer of mechanical energy.

Math Link

Giving and Taking

Think about the Explore Activity. How much work did the lever do when it lifted the weights? You need to know how high the weights were lifted and the force required to lift them. This force is their weight. Calculate this work from the data in your ***Activity Log*** on page 17.

Now, calculate the work done by you when you pulled down on the scale to lift the weight. Calculate this for each distance from the fulcrum using the data in your ***Activity Log.*** The work you did each time should be about the same and about equal to the work the lever did on the weight.

The work you did is the amount of energy you put into the lever. It was about the same amount each time. Which two things were different in each case? When one got larger, what did the other do? To keep the amount of work constant when force increases, distance must decrease.

To get a better understanding of how this happens, complete the table shown below. You will find a copy of this table in your ***Activity Log*** on page 19.

force	distance	work
1	12	12
2	?	12
?	?	12
?	?	12
?	?	12
?	?	12

The reason levers are so useful is precisely because they allow us to do a certain amount of work with less force or less distance (but not both) than could be done without the lever. A good example of how useful a lever can be is a bottle opener when your bottle doesn't have a screw-top cap. It takes a lot of force to push the cap off the bottle, but it doesn't have to move very far. This is an ideal job for a lever. You want to use the lever to change the force to a smaller amount, but in doing so you must move it over a larger distance.

Engineers group levers into three classes. The pictures on this page show examples of levers in each class—first, second, and third. The lever you made in the Explore Activity and the seesaw are first-class levers. Do the Try This Activity below to investigate how second- and third-class levers are different.

Scissors are formed from a pair of first-class levers.

If a fish takes the bait, the weight of the fish will be farther from the fulcrum than the hand of the person catching it.

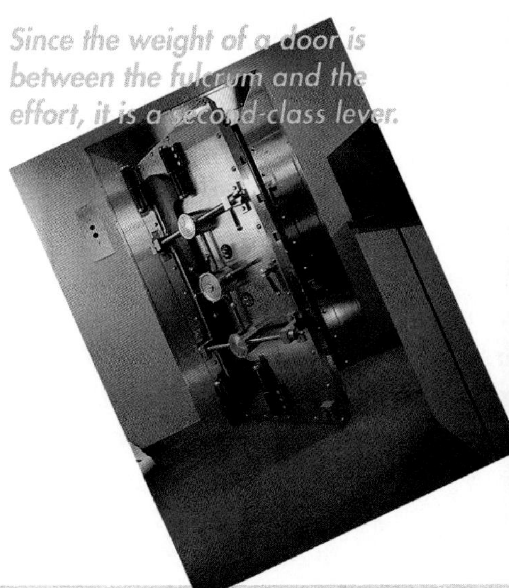

Since the weight of a door is between the fulcrum and the effort, it is a second-class lever.

TRY THIS Activity!

Hoist Again

A first-class lever has the fulcrum between the effort and the resistance. How does changing the position of the fulcrum in relation to the effort and the resistance make a lever act differently?

What You Need

spring scale
meterstick
ring stand
small piece of string
large washers
large paper clip
***Activity Log* page 20**

Using the same equipment as in the Explore Activity, investigate second- and third-class levers. Make these classes of levers by hanging the meterstick with the string at the end, as shown in the picture. Test how much force is needed to move the resistance. Vary the lengths of the effort and resistance arms.

Can you make a lever that causes the resistance to move more than the effort? Can you make a lever that changes the direction of the effort force? Write the answers and explanations in your ***Activity Log.***

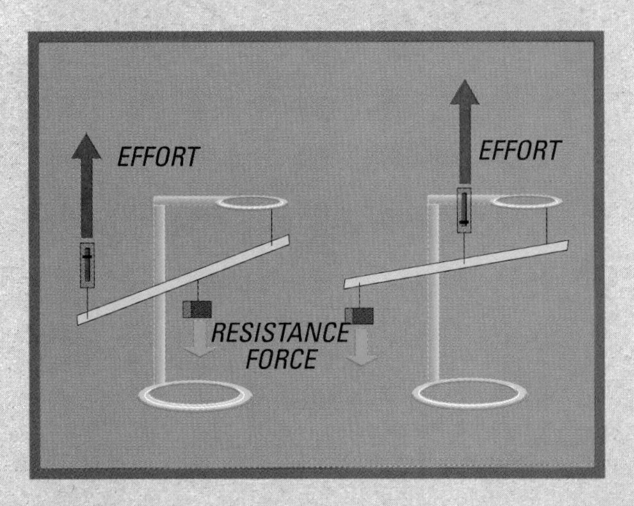

The fulcrum of an oar is between the effort and the resistance. A rowboat is one way to travel first class.

In **first-class levers** such as seesaws, boat oars, and crowbars, the fulcrum is located between the effort and resistance forces. The effort and resistance forces move in opposite directions in first-class levers. When you sit on a seesaw, you exert a force that pushes your partner upward. In the Explore Activity, you used a first-class lever to change the direction and magnitude, or strength, of a force. When you pulled down on the effort arm, the resistance arm rose up. When you moved the resistance toward the fulcrum, the magnitude of the resistance force decreased.

In **second-class levers,** such as the wheelbarrow, the fulcrum is located at one end and the effort force at the other. The resistance force is in between. Think about cracking a nut. Is the direction of the force you apply as you squeeze the nutcracker changed by the lever? If you answered "no," you are correct. Second-class levers do not change the direction of a force. Is it easier to crack a nut with a nutcracker than with your bare hands? Yes, second-class levers change the magnitude of the force you apply.

When you lift a wheelbarrow to empty or carry a load of wood, the wheelbarrow is a second-class lever. The wheel is the fulcrum. The load exerts the resistance force. Who exerts the effort force? When you use a bottle opener, the fulcrum rests on top of the bottle cap, and the resistance arm is under the lip of the cap. The effort arm is the handle. You exert the effort force to overcome the resistance force of the bottle cap. The cap comes off.

A wheelbarrow like this one is a good example of a second-class lever.

Have you ever used a shovel? You place the blade of a shovel into something like dirt or snow that you want to lift. To lift the load, you pull up on the handle of the shovel. The wrist of the hand at the top of the shovel acts as the fulcrum. The load is farther from the fulcrum than your other hand is.

Levers such as shovels, in which the effort force is between the fulcrum and the resistance force, are **third-class levers.** Like their second-class counterparts, third-class levers change the magnitude but not the direction of a force. Your forearm is a third-class lever. Can you identify the parts of this lever? If the elbow is the fulcrum, where are the resistance force and the effort force?

Since a shovel is a third-class lever, this man is exerting a force that is greater than the weight of the snow in the blade.

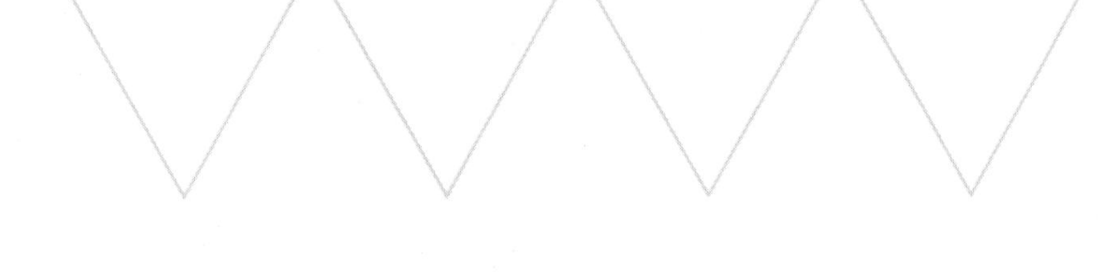

The first wheelbarrow was invented about 1,000 years ago in China. Its wheel was in the middle rather than in the front. People balanced the loads they carried in these wheelbarrows in front of and behind the wheel. The person who drove the wheelbarrow didn't have to do any lifting. He or she just steadied the wheelbarrow and guided it. The ancient Chinese hitched horses to some wheelbarrows, which meant that the horse, rather than the driver, supplied the effort force. A few people attached sails to wheelbarrows, using wind as the effort force. The next time you use a wheelbarrow, you can tell your family or friends that you are using an ancient Chinese invention.

This wheelbarrow could have been called a wheel seesaw.

Minds On! Look back at the pictures at the beginning of this lesson. Identify the class of each lever shown. List the levers and their classes in your ***Activity Log*** on page 21. ●

You use machines because they give you an advantage. The term **mechanical advantage** refers to how much force is increased by using a machine. The smaller the mechanical advantage, the greater the force you must supply to do work. The greater the mechanical advantage, the more work is done by the machine (and less by you)! The mechanical advantage of a lever can be determined by finding the ratio of the effort arm to the resistance arm.

$$Mechanical\ Advantage = Effort\ Arm \div Resistance\ Arm$$

$$MA = EA \div RA$$

Suppose you lift a nail from a board using a claw hammer. The resistance arm is 0.02 meter, and the effort arm is 0.25 meter. The mechanical advantage would be

$$MA = EA \div RA$$

$$MA = 0.25 \text{ m} \div 0.02 \text{ m}$$

$$MA = 12.5$$

You would be able to exert 12.5 times as much force using the claw hammer as you would without it.

Using a hammer as a first-class lever for removing a nail can give you a mechanical advantage.

Think about using a claw hammer to remove a nail from a board. In the process of removing the nail from the board, the surface of the claw hammer touches the surfaces of the nail and the board. Your hand grips the handle of the hammer. In Lesson 1 you learned about the force of friction that opposes motion between two surfaces touching each other. When a nail is lifted from a board, friction between the wood and the nail causes the nail to become warmer. Some of the mechanical energy from the effort force is changed into heat.

We sometimes say that energy is "lost" to heat when kinetic energy, such as the energy you use when removing a nail from a board, encounters friction. However, remember the law of conservation of energy? It states that energy is never really lost. The total amount of energy in a system remains the same. We sometimes talk as though energy is lost; however, what really happens is that energy changes to a form that doesn't help us do work.

Thermal energy isn't useful to you when you remove a nail from a board. The law of conservation of energy means, in part, that a machine can never do more work than you put into it. It can, however, produce a greater force. The hammer can't make up for the energy "lost" to heat. You have to make up for it. In the real world, you must put more work into the hammer than the hammer does work for you. This lowers the mechanical advantage of the hammer. The actual mechanical advantage would be the ratio between the resistance force and the effort force.

$$\text{Actual Mechanical Advantage} = \text{Resistance Force} \div \text{Effort Force}$$

$$AMA = R \div E$$

The actual mechanical advantage is always less than the mechanical advantage.

The actual mechanical advantage of this boat oar could be found by dividing the resistance force of the water against the blade by the effort force applied by the rower's hand and arm on the oar handle.

An oiled chain is more efficient than a rusty one.

In Lesson 3 you learned some ways in which engineers are trying to make more efficient cars that burn less gasoline. You know that modern, streamlined cars overcome friction caused by wind drag more easily than cars of the 1950s. In the Lesson 1 Explore Activity, you learned some other ways that friction can be reduced. Lubrication was one of the methods you learned about that reduces friction. Perhaps you have oiled a bicycle chain or even a squeaky door hinge. Sanding and waxing reduce friction, too. Maybe you've watched the Winter Olympics on television every four years. Champion and other downhill skiers wax their skis to get the right amount of friction between skis and snow.

Remember that work is the product of a force and a distance. ($W = F \times d$) Think of the work you put into a simple machine, such as a jack, as input. The work the jack puts into lifting the car is output. When you oil a bicycle chain, attach wheels to a go-cart, or sand a ramp, you increase the efficiency of a machine.

Efficiency measures how much work a machine can do compared to how much work must be put into the machine. Efficiency is a percentage. You obtain the efficiency of a machine by dividing the work output by the work input and multiplying this ratio by 100.

$$\textit{Efficiency}\ (\%) = (\textit{Work Output} \div \textit{Work Input}) \times 100$$

Minds On! Suppose that, using a lever, you exert an effort force of 50 newtons for 20 centimeters. The resistance, which weighs 90 newtons, moves 10 centimeters. Use the formula *Percent Efficiency* = (*Work Output* ÷ *Work Input*) × 100 to calculate the efficiency of the lever. Hint: Remember to change centimeters to meters, since joules are actually newton-meters. ●

Many machines don't approach the level of efficiency you calculated for the lever in the Minds On. For example, a typical automobile engine has an efficiency of only about 30 percent. Part of the reason is because the energy input is not mechanical work, but thermal energy that explodes in cylinders. Efficiencies can never be 100 percent when converting thermal energy to mechanical energy, even if there is no friction. The efficiency of any machine is always less than 100 percent.

Look at the pictures on this page. Each picture shows a wheel and axle, or rotating lever. Notice that these simple machines are made up of two circles, an axle (the smaller circle) and the wheel (the larger one). It's not always easy to tell that the wheel is a circle.

For example, look at the pencil sharpener drawn from two angles. The crank (Figure 1) is actually the wheel. If you want to sharpen a pencil, you apply an effort force to the wheel (you turn the crank). Figure 2 shows the smaller wheel, the axle. Inside the pencil sharpener, the axle connects to blades. When you exert a force on the wheel by turning the crank, the axle rotates. The blades cut wood from the pencil.

Suppose you sharpened a pencil with a hand-held blade. How would the effort force compare in magnitude with the force that you exert on a pencil sharpener? (The effort force would be much greater if you were sharpening the pencil by hand.) A pencil sharpener allows you to sharpen your pencil with less force. But you can't get something for nothing. What do you give up? To find out, sharpen a pencil using a pencil sharpener. Watch the crank and axle turn as you sharpen the pencil. Notice how much farther the crank turns than the axle with each revolution. You give up distance.

Wheels and axles can be used to maximize horsepower.

A pencil sharpener is an example of a wheel and axle. The crank is the wheel. What do you see on the axle? How is a pencil sharpened using this machine?

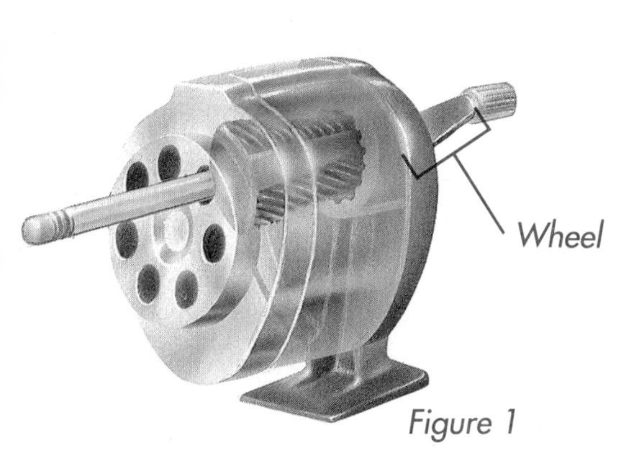

Figure 1

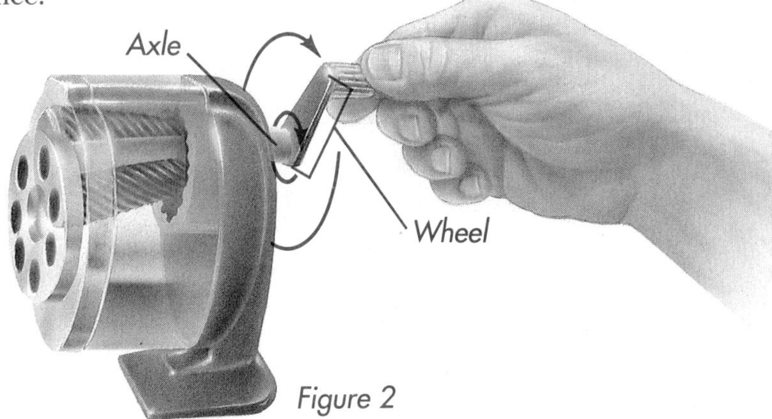

Figure 2

Minds On! Look at the doorknob and the screwdriver pictured on this page. Why are the doorknob and the screwdriver examples of wheels and axles? Write your explanation in your ***Activity Log*** on page 22. ●

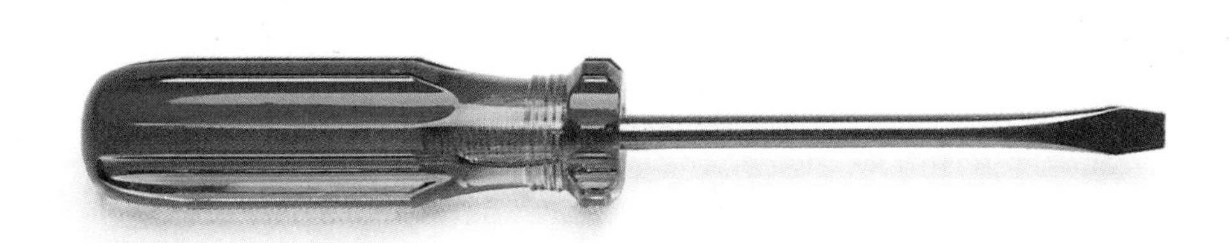

Have you ever raised a flag up a pole? The device at the top of the pole is a pulley, a simple machine that can change the magnitude and direction of a force. Like the wheel and axle, the pulley is a kind of lever. Think about raising a flag. You attach the flag to a rope. (The rope circles around the pulley at the top of the flagpole.) You pull down on the rope. The flag goes up. The pulley changes the direction of the effort force, but not, in this case, the magnitude.

A pulley system can also be used to change the magnitude of the effort force. Do the Try This Activity below to see how pulleys work.

TRY THIS

Activity!

Pull a Little; Lift a Lot?

How does changing the position of a pulley change the force needed to lift a load? Does adding a pulley make any difference?

What You Need

pulley kit (including cord), spring scale, laboratory mass, ringstand, *Activity Log* page 23

Using the spring scale, weigh the laboratory mass. Record the reading in your ***Activity Log.***

Set up the pulley system as shown in Figure 1. Use the spring scale to lift the weight. Record how much effort force it took to lift the weight. How did the single pulley affect the magnitude of the effort force? How did it affect the direction of the effort force? Write the answers in your ***Activity Log.***

Now, set up the pulley as shown in Figure 2. Repeat the investigation you did with the pulley in Figure 1. Make sure to record the data.

Repeat the investigation using the setup shown in Figure 3. How much force did you use to lift the weight? How did this pulley system affect the effort and resistance forces? Record your answers in your ***Activity Log.***

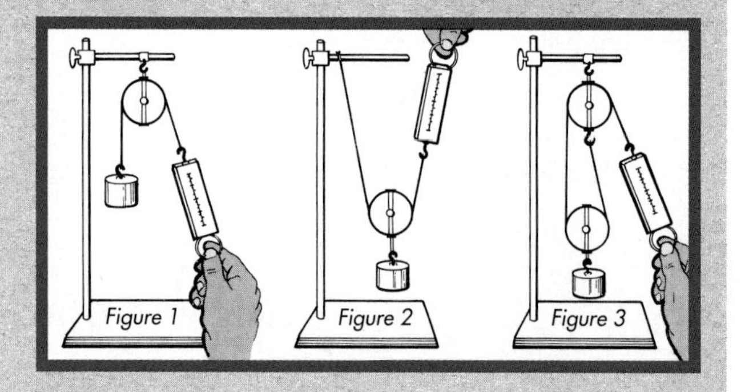

How did the effort force change as pulleys were added in the Try This Activity? Pulling down on a single pulley to lift something upward does not decrease the effort force. Using a two-pulley system for pulling downward to lift something upward decreases the effort force considerably. Compare the effort force needed when you pulled downward using one pulley and two pulleys. Can you determine any mathematical relationship? In a pulley system without friction, the effort force would be equal to the resistance force divided by the number of ropes that support the resistance. For example, if an effort force of 12 newtons is required to lift a weight using one rope to support it, the effort force using two ropes supporting the weight would be

Effort Force = Resistance Force ÷ # of Ropes

12 newtons ÷ 2 = 6 newtons

You probably didn't have measurements that were exactly equal to the resistance force divided by the number of supporting ropes. Some of your effort was used to overcome the opposition of friction.

Machines Inside and Outside

The pulley systems you have investigated so far can't be used to lift a heavy weight. Look at the picture on this page. It shows a pulley system lifting a person. Fixed and movable pulleys are combined in this block-and-tackle system. Trace the four strands of rope used in this system. All four strands carry the effort force that the person applies. A similar system can be used by car mechanics to lift a car engine. The mechanic pulls down. The engine rises up. In a world without friction, this pulley system would quadruple the effort force. The following Career Feature explains more about how automobile mechanics use machines.

Several pulleys can be combined to make a block and tackle with a large mechanical advantage.

Automobile Mechanic

Since she was a little girl, Susan Chan has been intrigued with machines such as bicycles and automobiles. As an automobile mechanic who specializes in rebuilding transmissions, Susan has put her lifelong interest and mechanical aptitude to work.

Susan works in the transmission department of a large automotive repair shop in Dallas, Texas. Her job involves adjusting, repairing, and maintaining complex parts of automobiles such as gear trains, hydraulic pumps, and other parts of the automatic transmission system. Susan knows a lot about simple machines!

Taking a high school course in automobile repair procedures gave Susan a sense of what a mechanic does and confirmed her hunch that she would enjoy the work. High school courses in mathematics, physics, and chemistry gave Susan the background knowledge to understand what she later observed on the job. Working in a gas station after school and during the summers gave Susan a head start in her career.

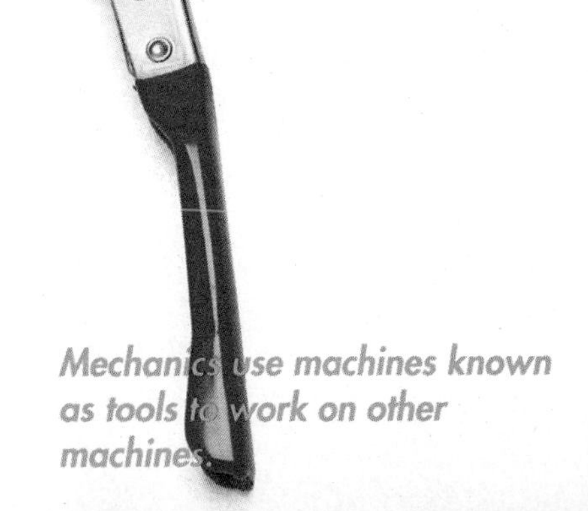

Mechanics use machines known as tools to work on other machines.

A transmission specialist like Susan spends about four years in either a formal apprenticeship program or on-the-job training and another year or two acquiring more practical experience before she or he becomes eligible for certification by the National Institute for Automotive Service Excellence.

For more information, write to:
Automotive Service Industry Association
444 North Michigan Avenue
Chicago, IL 60611

An automobile mechanic like Susan Chan may use a pulley system to lift a heavy engine or a transmission. Pulley systems can also increase your effort force enough for you to lift yourself. The following Focus on the Environment explains how biologists use a pulley system to get a lift.

Susan skillfully uses machines to repair an automobile.

Focus on Environment

In the Trees

From a tram suspended some 40 meters (about 131 feet) above the ground, biologist Don Perry identifies and observes the rich variety of plants and animals that live in the tropical rain forests of Costa Rica. The tram is an enormous pulley system. In a typical workday, Perry bobs across the rain forest along a 260-meter (about 853-foot) stainless steel cable strung between two trees. He can move up and down as well as across. His control? A joy stick designed to control a model airplane.

A human can be one of the forest's better climbers by using pulleys.

"What intrigued me," Perry says about the rain forest, "was that there would be snakes and lizards and ants and wasps and a whole community of life up there in the trees." There is an urgency to Perry's work because the rain forest is disappearing. Perhaps you already know something about this problem. A change in one part of an ecosystem, such as a rain forest, affects the whole. Humans are changing the rain forests by burning them to clear land to raise cattle. In the process, valuable medicinal plants that can help fight cancer, leukemia, and muscular and heart disease are being destroyed. By logging rain forest trees faster than the trees can regenerate, people are destroying the habitats of some of the most diverse animal species in the world.

Tropical rain forest habitats are home to many species of living organisms. Scientists are discovering new species as they probe rain forests.

Perry's "Automated Web" of ropes and pulleys will speed rain forest biologists' work enormously as they explore the three topmost layers—understory, canopy, and emergent layer. These biologists can then catalogue the organisms they find there. The device is not problem-free, however. Like a car, the Automated Web runs by burning gasoline. The gas tank has run out twice, leaving Perry stranded in the trees. Stranded, that is, until he used a climbing rope to rappel, or drop himself, 40 meters (131 feet) to the forest floor!

Minds On! Imagine you are a newspaper reporter assigned to interview a rain forest biologist like Don Perry. Working in groups of three or four, generate a list of questions and research your answers. Then, as a group, write a short news interview. ●

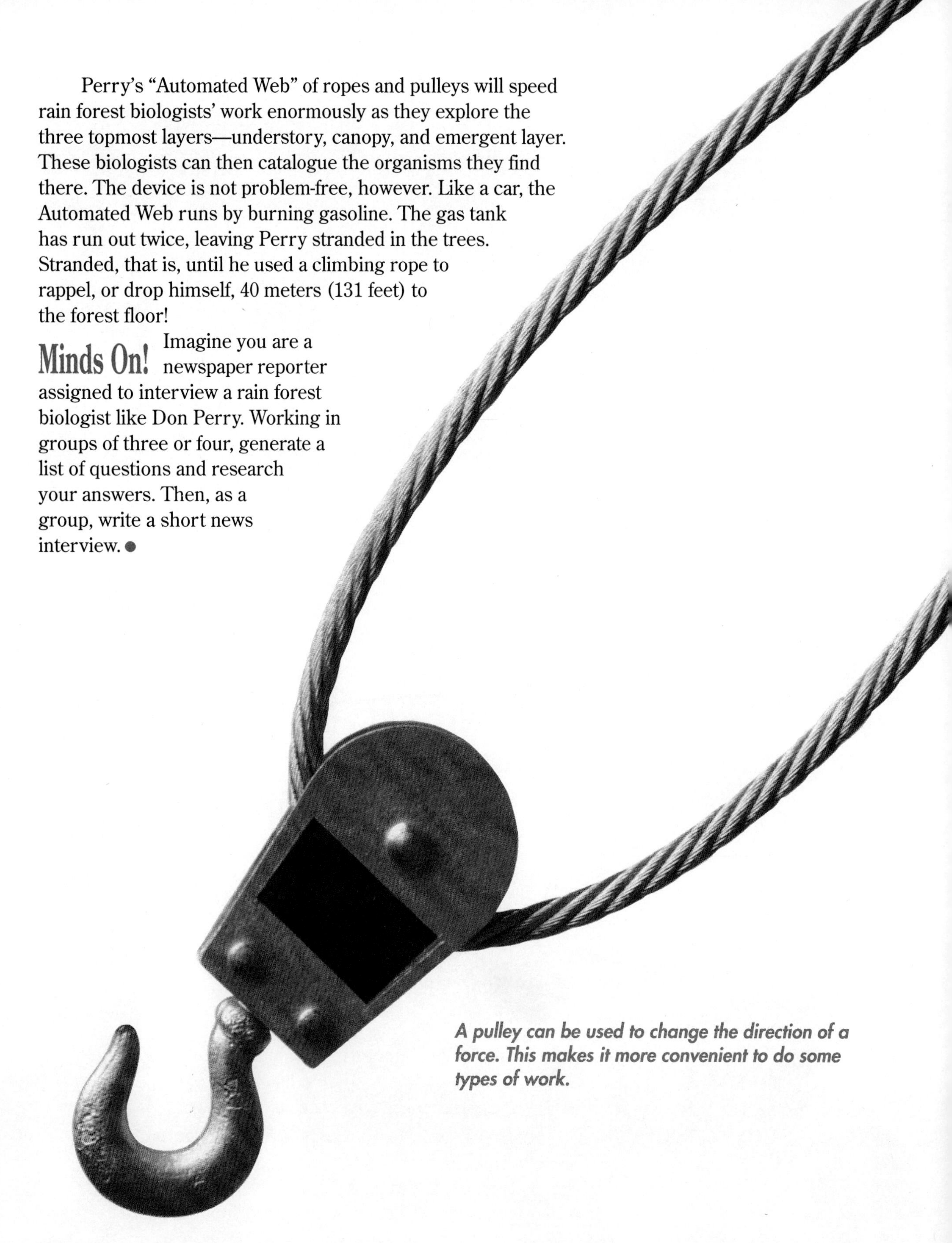

A pulley can be used to change the direction of a force. This makes it more convenient to do some types of work.

Sum It Up

Levers are simple machines that can change the magnitude and direction of a force. The law of conservation of energy determines how much the force changes when the distance changes. The relationship between force and distance is the same for first-, second-, and third-class levers. These levers differ by the position of the fulcrum in relation to the effort and the resistance. The mechanical advantage, actual mechanical advantage, and efficiency for a lever or a system of levers can be calculated from force and distance measurements.

Wheel and axles and pulleys are also levers. Wheel-and-axle systems change force by changing distance like other levers do. Pulleys are useful in systems like a flagpole where the pulley changes the direction of a force. A system of pulleys known as a block and tackle can also decrease the effort force to enable people to lift a heavy motor or even themselves.

Using Vocabulary

effort arm, effort force, first-class levers, fulcrum, lever, mechanical advantage, resistance arm, resistance force, second-class levers, simple machines, third-class levers

Write a story about two people fishing in a rowboat. While waiting for a fish to take the bait, they are snacking on unshelled pecans. Use each of the words from the list in your story.

Critical Thinking

1. The handles of a hedge clipper are longer than its blades. The handles of scissors are shorter than the blades. Explain why these machines are made this way.
2. Describe two ways to use a single pulley to lift a bucket of nails from the ground to a second-floor scaffold.
3. If Noah wanted to use a steel fence post as a lever to push a large rock, would it be better to use the post as a first-, second-, or third-class lever? Explain.
4. Juyong has a 2-meter wooden pole with a hook on one end. She wants to use it as a lever to carry a piece of luggage, with her shoulder as the fulcrum. If the piece of luggage weighs 200 newtons, and she can lift it with an effort force of 67 newtons, what is the actual mechanical advantage of the lever?
5. Jennifer uses a pulley system to lift an 800-newton transmission from the floor to a height of two meters. If she pulls with a force of 250 newtons for a distance of eight meters, what is the efficiency of the pulley system?

Theme T SYSTEMS and INTERACTIONS

HOW DO *Inclined Planes* HELP *People do Work?*

How can a long, hard climb be made easier, but not shorter? If you were climbing a rope, why would you fear a lumberjack with an axe more than a carpenter with a hammer? A simple machine can be used to make or break a climb.

Mountain climbers use ropes and protective gear when making a climb. Sometimes they climb up sheer rock faces, but other times their climb is more gradual. Look at the photograph of the climber on this page. What would it be like to climb on this surface? Notice the slant of the rock face. How would it be easier to climb this type of rock than one that was at a steeper angle? On the other hand, think about the time it would take to make both climbs if the distance to the top was the same. Which would you rather climb?

Ramps and slopes are inclined planes.

Minds On! What are the advantages of inclined planes? When are they most useful? Use the pictures on this page to answer the questions. Record your answers in your ***Activity Log*** on page 24. ●

EXPLORE

Activity!

Let It Rip!

You know all simple machines help people do work. Some simple machines, such as a seesaw and a block and tackle, can change the magnitude and the direction of the force that you apply. Other machines, such as nutcrackers, change the magnitude but not the direction of an applied force. In this activity you'll investigate how a simple machine called an inclined plane affects the effort force.

What You Need

flat board at least ½ m long
spring scale
string
book
meterstick
***Activity Log* pages 25–26**

What To Do

1 Using the spring scale, find the weight of your resistance (book). Record this information in your ***Activity Log*** table.

2 Calculate the amount of work needed to lift the book directly for each height in the table, without using the inclined plane. Record this information under the column "Work output."

Height of plane (m)	Resistance force (N)	Work output (J)	Effort force (N)	Length of plane (m)	Work input (J)	Efficiency
0.05 m						
0.10 m						
0.15 m						
0.20 m						
0.25 m						

3 Set up an inclined plane to make a ramp as shown. Make sure that the end is raised 5 cm (0.05 m). Use the spring scale to pull the book up the incline. Pull at a slow, steady speed. Record the effort force needed to lift the book by pulling it up the ramp.

4 Measure the length of the ramp. Then, calculate the amount of work required to pull the book up the ramp.

5 Repeat this procedure for ramp heights of 0.10 m, 0.15 m, 0.20 m, and 0.25 m.

What Happened?

1. Which took more force, lifting the book straight up or using the ramp?
2. You experimented with ramps of five different heights. How high was the ramp when the smallest amount of force was required?
3. Did it require less work to pull the book up the ramp than it did to lift the book to the same height directly?

What Now?

1. Calculate the efficiency of the inclined plane by dividing the work output (that the machine did) by the work input (that you did). How efficient was your machine?
2. Why was the work output different from the work input?
3. What do you think would happen to the effort force if you had a ramp that was twice as long as the longest one that you used? To the input work? Work with another group to try it.
4. Think about how you could improve the efficiency of your ramp machine. Try out your ideas.

Three Machines for Sliding Force

An inclined plane increases the distance you move an object. Less force is needed than lifting it vertically to the same height.

In the Explore Activity, you experimented to see how an inclined plane affects the force that you exert to do the work of lifting a weight. An **inclined plane** is a straight, slanted surface. You found that the inclined plane decreases the magnitude of the force needed to lift the weight. Do you do less work to accomplish the same task with an inclined plane than without one? No. Look at the data you recorded in your ***Activity Log.*** The amount of work done to lift the weight directly and using the ramp was just (or just about) the same.

Minds On! In a world without friction, the amount of work done with and without the ramp would be *exactly* the same. Use what you observed in the Explore Activity and what you know about friction to explain why. Write the answer in your ***Activity Log*** on page 27. ●

Gravity

Think about what happened when you increased the height of the inclined plane. The magnitude of the effort force required to lift the weight also increased. In Lesson 4 you learned that mechanical advantage measures how much force you gain by using a machine. Remember that the actual mechanical advantage of a machine is less in the "real world." Why?

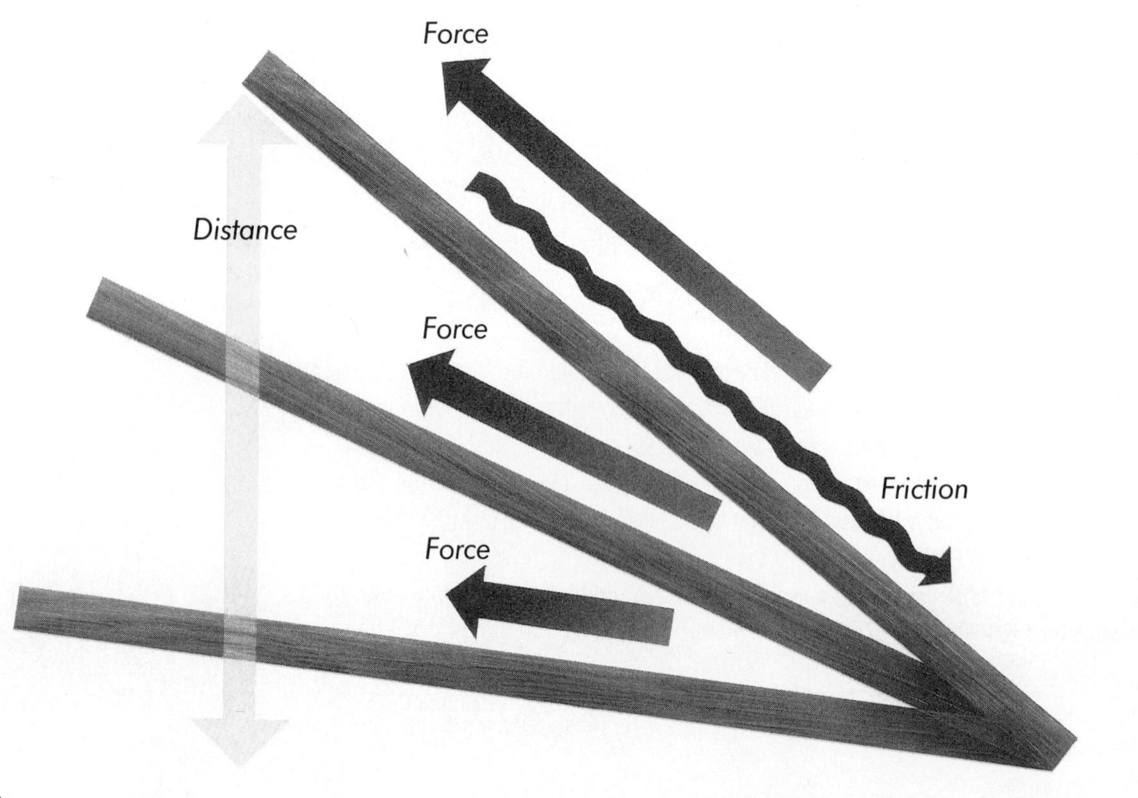

Minds On! You can calculate the actual mechanical advantage of an inclined plane such as a ramp in the same way you calculated the advantage of a lever: by dividing the resistance force by the effort force. $(AMA = R \div E)$

Use the data you recorded in the Explore Activity to calculate the actual mechanical advantages of the ramps with heights of 0.05 m, 0.10 m, 0.20 m, and 0.25 m. Which ramp had the greatest mechanical advantage? Explain why in your ***Activity Log*** on page 28. ●

In Lesson 4 you learned that *efficiency* measures how much work you get from a machine compared to how much work you put into the machine. Look at the table you made and find the efficiency of the inclined planes that you used in the Explore Activity. Did the efficiency exceed 100 percent? No. You know why. If a ramp had an efficiency of, say, 110 percent, you would be getting something for nothing. Such a ramp would do more work than you put into it. For that to occur, the ramp would have to somehow increase the energy that you transfer to it when you use it to do work. That can't happen. The amount of energy in a system can neither increase nor decrease. The amount of energy in a system stays the same.

Perhaps you thought of a way to make your inclined plane more efficient by reducing friction between the surfaces of the ramp and the weight. In Lesson 1 you learned that lubrication and rolling reduce friction. If you attached wheels to the weight, you reduced friction. If you oiled the ramp, you reduced friction.

Reducing the friction between the resistance and the ramp is a good way to increase efficiency.

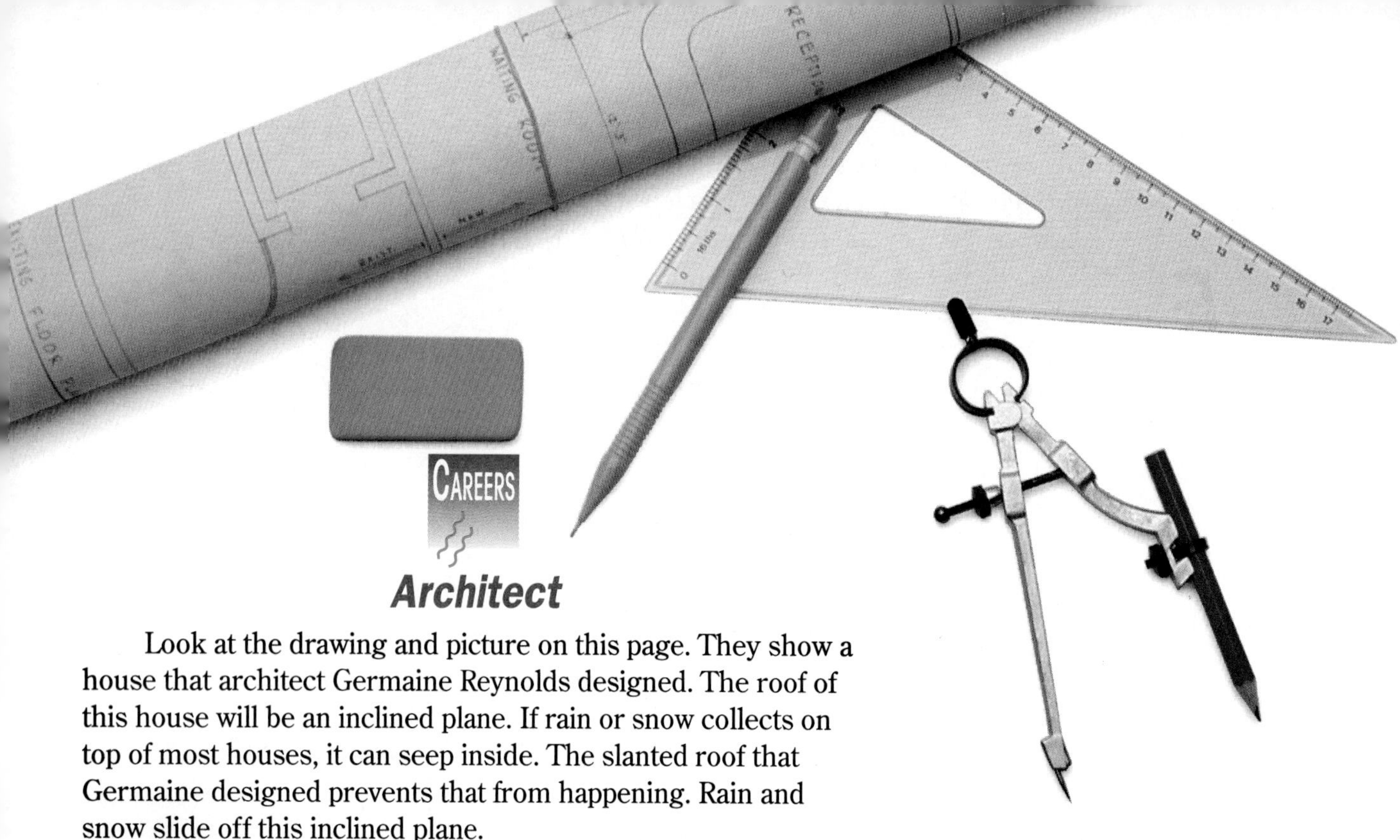

Architect

Look at the drawing and picture on this page. They show a house that architect Germaine Reynolds designed. The roof of this house will be an inclined plane. If rain or snow collects on top of most houses, it can seep inside. The slanted roof that Germaine designed prevents that from happening. Rain and snow slide off this inclined plane.

In high school, Germaine took college preparatory courses in mathematics, physics, art, English, and other disciplines. Although he doesn't consider himself a great artist, Germaine does have the ability to visualize spatial relationships, which is a key to his architectural success. Germaine majored in communications at the state university he attended, partly because he knew that an architect must be able to communicate effectively in words and writing, as well as drawing. After college, he entered a four-year master of architecture program.

If you enjoy freehand drawing and working with people, then perhaps a career in architecture is for you. You can receive more information by writing to:

American Institute of Architects
1735 New York Avenue, NW
Washington, DC 20009

You know there are two basic kinds of simple machines, the lever and the inclined plane. Two variations, or different kinds, of levers are wheel and axles and pulleys. A wheel and axle, such as a pencil sharpener, is a rotating lever. A pulley is a lever that rotates around a fixed point, such as the top of a flagpole.

Inclined planes have variations, too—the screw and the wedge. Do the next Try This Activity to find out about one of these variations.

The inclined plane is found in many architectural designs.

TRY THIS Activity!

It's a Wrap

What simple machine holds doorknobs to doors, bicycle seats to bicycle frames, and STOP signs to posts? The screw. The screw is a kind of inclined plane. It's easy to see how if you make a model.

What You Need

paper, pencil, scissors

Cut a piece of paper along the diagonal to form a triangle. Turn the triangle so that the longest side, or hypotenuse, is on the right. Does the triangle look familiar? As you see, it looks just like an inclined plane.

Wrap the paper triangle around a pencil. You have made a model of a screw.

The handrail on this staircase is a screw.

An inclined plane decreases the effort force when a vehicle uses a ramp.

Have you ever slid onto the ground or into a swimming pool on a spiral slide? Think about the shape of the slide. It's like a screw. A **screw** is an inclined plane wrapped around a cylinder. Suppose you could "unravel" the slide. What shape would you find? Think about how the spiral slide, the model of a screw, and the ramps that you built in the Explore Activity are the same. Now think about using each of these inclined planes to do work.

You know that you used the ramps to do work in the Explore Activity. You measured the work you did. Entrance ramps and exit ramps along highways help the engines of vehicles do work. For a vehicle to make a climb, the engine must do work by pulling the weight of the vehicle and its passengers up the height of the climb. Using a straight ramp or a spiral ramp allows the engine to pull with less force. Since the ramp is longer than the height of the climb, the engine has to pull for a longer distance when the vehicle uses the ramp. The engine doesn't do less work, but it does pull with less force.

Think about doing work by turning a screw. Suppose two screws have the same diameter and length, but one screw has more threads than the other. The screw with more threads has less distance between each thread. The screw with fewer threads requires more force to turn it one revolution. The screw with more threads needs less force to turn it one revolution, but it doesn't penetrate as far. Less force also results in less work being done for each revolution of the screw.

Suppose you had a giant machine like the one pictured on this page. Construction crews have used this screwlike machine to bore through solid rock. This machine was used to build a tunnel under the English Channel to connect England and France.

You know that one variation of the inclined plane is the screw. A screw is made up of an inclined plane and a cylinder. Think about the model you made. The triangular piece of paper was the inclined plane. What was the cylinder?

Another variation of the inclined plane, the **wedge,** has either one or two sloping sides. Look at the wedges shown on this page. Note the inclined planes on each wedge. Think about how people use the wedges shown to do work. People use axes to split wood, blades to cut paper, and plows to turn soil.

This machine has helped to connect England to France.

Some wedges are used to plow the soil.

TRY THIS **Activity!**

Inspecting a Doorstop

What height slope makes the best doorstop?

What You Need

doorstop, small piece of paper, scissors, *Activity Log* page 29

Perhaps a doorstop holds the door of your classroom open. Is it attached to the door? How does it work? Look at a different kind of doorstop. Hold it by its narrow end and slide it so the opposite end is against a wall. What do you notice? Now take your paper and cut it diagonally from corner to corner. How does the slope of the paper you cut differ from the slope of the slanted side of the doorstop? Which would make the better doorstop? Why? Write your answers in your *Activity Log.*

People use wedges to increase the magnitude of a force exerted between two objects. Suppose you tried to split a log with your bare hands. It couldn't be done, even if you had a black belt in karate. Neither could you cut paper with your fingers. Could you neatly cut a piece of pie using only your finger? Another wedge known as a knife makes clean cuts. You couldn't cut through an apple with your gums. You need your teeth to act as a wedge. It takes less effort force to break a piece from an apple using wedge-shaped incisors.

Both sides of this wedge are inclined planes.

Think about splitting a log. In what direction do you move the axe? You exert a downward force through an object such as a log with a wedge. You could also lift something upward using a wedge. You would apply the effort force on the wedge at a horizontal direction while the object was lifted vertically.

Simple machines are found in the bodies of many organisms. Remember in Lesson 4 you learned that your forearm is a third-class lever. Have you ever considered how different your life would be if you didn't use simple machines to do work? The following Literature Link focuses on life on a planet without machines.

Literature Link

The Guardian of Isis

"All through the winter Jody has thought about the river, falling from the heights and pounding the basin at the foot of the Cascades with such force that the rocks trembled and the air was filled with spray and thunder. So much power, all going to waste. . . .

—If you could make the water do *something as it fell from the high country into the Valley, something like turning a wheel, then you could do . . . do what? What use was a wheel turning and turning by itself, way up at the head of the valley? Jody felt as if he was on the edge of a tremendous discovery."*

In the science-fiction novel *The Guardian of Isis* by Monica Hughes, Jody N'Kumo is growing up in a community on the planet Isis that has no machines. What is he about to invent in the excerpt above? Will this invention industrialize Jody's valley? If it does, what kinds of machinery do you think a small community that has never known machines would develop first? Working in a small group, try to predict the career of Jody N'Kumo, Inventor and Industrializer of the planet Isis. If you've already read *The Guardian of Isis,* also consider this question: Why do you think the author chose not to answer these questions in the novel?

Artistic and Automated Inclines

You know that architects like Germaine Reynolds often make use of inclined planes in their designs. Look at the picture on this page. It shows the Solomon R. Guggenheim Museum in New York City. You could think of this art museum as one enormous simple machine. Notice the spiral ramp, or grand ramp, at the center of the museum. In designing the museum, architect Frank Lloyd Wright chose as his starting point a seven-story inclined plane made of reinforced concrete.

The grand ramp contributes to the way the paintings found in the Guggenheim Museum are perceived. A visitor to the Guggenheim Museum can see all the paintings from any place on the ramp. The walls that the paintings are on are curved. As a person moves up or down the ramp, the patterns in the paintings appear to be different. If you've ever looked into a kaleidoscope, you have some idea of the combined effects of walking the ramp and looking at the paintings on the curved walls. As you turn a kaleidoscope, the shapes and patterns inside it combine in different patterns.

A seven-story inclined plane gives the Guggenheim Museum some interesting qualities.

Many places—like museums, airports, and shopping malls—where large crowds of people need to go from a lower floor to an upper floor, use escalators. An escalator uses two simple machines, the inclined plane and the pulley, to lift people from one level to another.

Study the diagram of the escalator on the next page. How is the escalator an inclined plane? Unlike an elevator, which lifts people directly, an escalator lifts people on an incline. In the Explore Activity, you learned that lifting a weight on an incline requires less force than lifting the same weight directly. An escalator gains a mechanical advantage by lifting people on an incline.

Look at the diagram again and notice that an escalator also has chains and gears. A gear is actually a pulley with teeth. The chain has a function similar to the rope on the flagpole. A rope transfers mechanical energy from you to the flag. The chain transfers the mechanical energy from the motor to the steps on the escalator and pulls the people upward.

Sum It Up

You are now well acquainted with both kinds of simple machines, levers and inclined planes, and their variations, wheel and axles, pulleys, screws, and wedges! These machines can be found in many systems where the magnitude or direction of a force is changed. Inclined planes, screws, and wedges permit you to use less effort force in machines like a ramp or axle. In an imaginary journey to the Guggenheim, you learned about the artistic use of inclined planes in buildings and museums. Buildings and museums also use inclined planes and pulleys to lift people on an escalator.

An escalator lifts people on an incline.

Using Vocabulary

inclined plane
screw
wedge

Describe some kind of work that is done using each of these simple machines.

Critical Thinking

1. Roads that go to the top of small hills are usually straight inclines, but roads to the top of larger hills and mountains are spiral-shaped. Use what you have learned about inclined planes and screws to explain this difference in design.
2. How could you use a wedge to lift a crate of fruit?
3. The bottom of a door at the entrance of a building is one meter above ground level. If the ramp from the ground to the door is five meters long, how much force, ideally, would it take to pull a wagon that weighs 750 N up the ramp?
4. Darnell exerted a force of 800 newtons to push Cindy in a wheelchair up the ramp referred to in question 3. If Cindy weighs 650 newtons and the wheelchair weighs 100 newtons, what is the efficiency of the inclined plane/wheel-and-axle system?
5. Unlike an escalator, an elevator does not use an inclined plane to gain a mechanical advantage. How could an elevator gain a mechanical advantage without using an inclined plane?

Look back at the machines illustrated at the beginning of this unit on pages 6 and 7. Could you tell when you first saw these items that they were machines?

You now know that a pencil sharpener, such as the one pictured on page 6, is a wheel and axle. This simple machine changes the magnitude of the force that you apply by turning the wheel, or crank. This turns an axle, which is connected to the wedges that cut wood from the pencil.

Minds On! At the beginning of this unit, you identified the items in your kitchen that are machines and listed them on page 1 in your ***Activity Log.***

Without referring to that list, go back to your kitchen. Again, identify the items that are machines. Record the machines you identify on page 30 in your ***Activity Log.*** Now, look back at the first list you made. Compare your original list with your final one. In your ***Activity Log,*** explain the differences between the two lists. How did the information learned in this unit affect the second list that you made? ●

Look at the wheel and axle on this bicycle. What forces are used in moving it up the inclined plane?

MACHINES are useful

Think about the different simple machines and how they help people do work. A lever such as a seesaw can change the magnitude and direction of a force. A wheel and axle, such as a screwdriver, can increase the force you apply. A movable pulley, in which several pulleys are combined, can strengthen an applied force tremendously, in addition to changing its direction. An inclined plane, such as a ramp, makes it easier for a person on roller skates or in a wheelchair to move into or out of a building.

Think about a bicycle. Like many machines you use, a bicycle is made up of several simple machines. Remember the simple machines found in an escalator? Look at the bicycle shown on this page. Identify the levers on the bicycle. What simple machines are the back wheel and pedals?

Suppose you ride a bicycle like the one shown. What happens when you push down on the pedals? The tire pushes against the road. The chain that connects the back wheel and pedals changes the direction of the force that you apply. You push down; the wheel moves around. The faster you pedal, the faster the wheel moves.

Many other machines besides a bicycle consist of two or more simple machines working together. A shovel is made up of a wedge, which you use to cut into the soil, and a lever, which you use to lift the soil. A push-type lawn mower is made up of many wedges (cutting blades) attached to a wheel and axle. A hand drill is made up of a wheel and axle and a screw (the bit that you attach to drill a hole).

Minds On! How many machines that are made of two or more simple machines can you identify in your school? Keep a running list for two days. Record the list in your ***Activity Log*** on page 31. Compare your list with the lists of other students. Then, working in pairs, identify the simple machines in each of these machines. ●

Normally a machine helps people do work. Musicians can work very hard in a performance, although they are said to be playing, not working, their instruments. Read the following Music Link to learn how simple machines work together in a piano.

Music/Art Link

Notable Levers

Have you ever watched someone playing the piano while the action and the strings were exposed? If you have, you observed the piano player pushing downward on the keys while the hammers of the piano pushed upward on the strings and the damper rose above the strings. How can one downward effort force result in moving two separate resistance forces upward?

The piano is a musical instrument that uses a combination of first-, second-, and third-class levers. The diagram below illustrates how a key action of a piano is constructed, along with the names of each part. When the piano player pushes downward on the key, the wippen raises upward. There's a fulcrum between the wippen and the key. Pushing downward on the key also raises the damper with a fulcrum between the key and the damper. The wippen pushes upward on the jack, which also pushes upward on the hammer. The wippen is located between the fulcrum and the jack. When the jack pushes upward on the hammer, the hammer raises and strikes the string. The jack is situated between the hammer and the fulcrum. The combined effect of all these levers transferring the effort force of the piano player is the vibration of the piano string and a musical note.

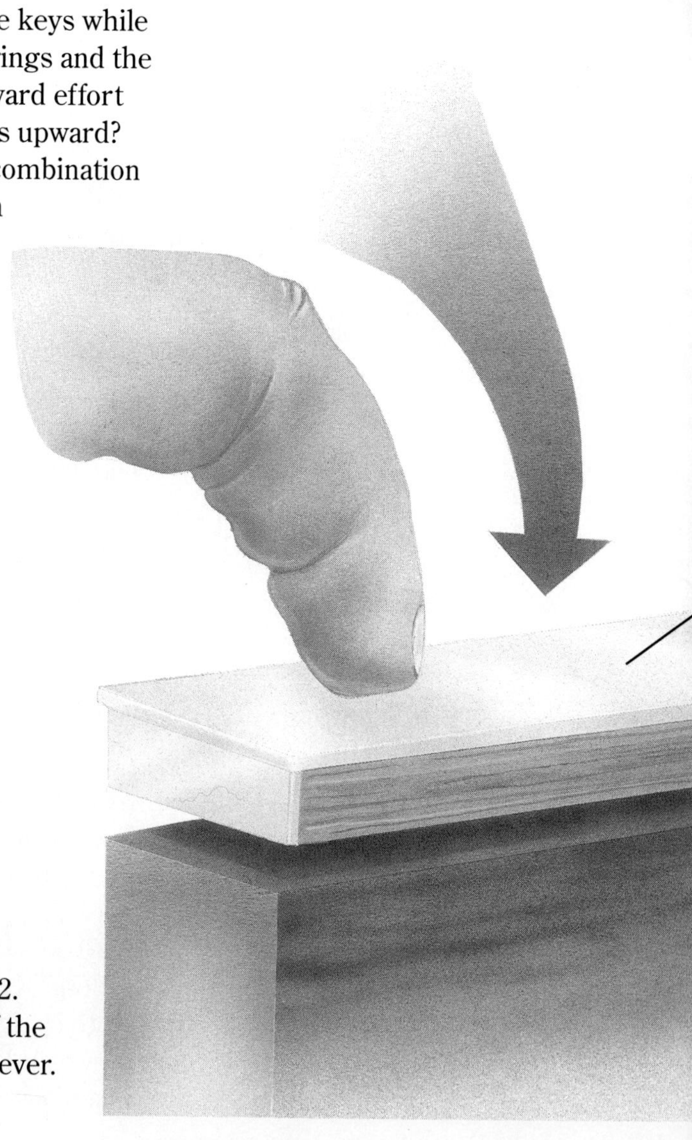

Study the diagram and the description of how a piano works. Try to determine the location of each lever used to play a single note. Label these levers in the diagram in your ***Activity Log*** on page 32. To be able to easily distinguish these levers, think of the name of a musician for whom you could name each lever. Add these names to each lever that you labeled.

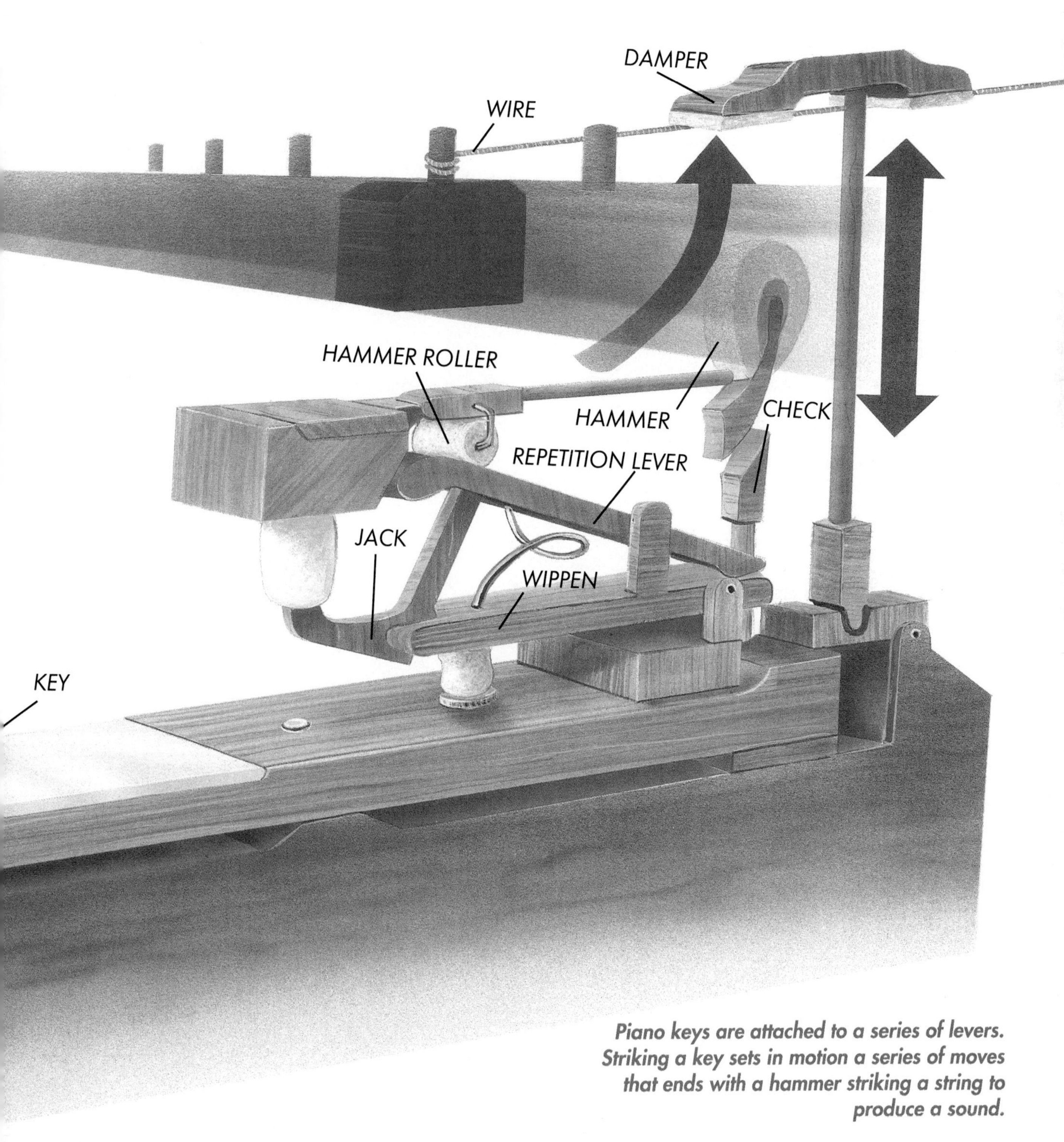

Piano keys are attached to a series of levers. Striking a key sets in motion a series of moves that ends with a hammer striking a string to produce a sound.

GLOSSARY

Use the pronunciation key below to help you decode, or read, the pronunciations.

Pronunciation Key

a	**a**t, b**a**d	d	**d**ear, so**d**a, ba**d**
ā	**a**pe, p**ai**n, d**ay**, br**ea**k	f	**f**ive, de**f**end, lea**f**, o**ff**, cou**gh**, ele**ph**ant
ä	f**a**ther, c**a**r, h**ea**rt	g	**g**ame, a**g**o, fo**g**, e**gg**
âr	c**are**, p**air**, b**ear**, th**eir**, wh**ere**	h	**h**at, a**h**ead
e	**e**nd, p**e**t, s**ai**d, h**ea**ven, fr**ie**nd	hw	**wh**ite, **wh**ether, **wh**ich
ē	**e**qual, m**e**, f**ee**t, t**ea**m, p**ie**ce, k**ey**	j	**j**oke, en**j**oy, **g**em, pa**g**e, e**dg**e
i	**i**t, b**i**g, **E**nglish, h**y**mn	k	**k**ite, ba**k**ery, see**k**, ta**ck**, **c**at
ī	**i**ce, f**i**ne, l**ie**, m**y**	l	**l**id, sai**l**or, fee**l**, ba**ll**, a**ll**ow
îr	**ear**, d**eer**, h**ere**, p**ier**ce	m	**m**an, fa**m**ily, drea**m**
o	**o**dd, h**o**t, w**a**tch	n	**n**ot, fi**n**al, pa**n**, k**n**ife
ō	**o**ld, **oa**t, t**oe**, l**ow**	ng	lo**ng**, si**ng**er, pi**n**k
ô	c**o**ffee, **a**ll, t**au**ght, l**aw**, f**ou**ght	p	**p**ail, re**p**air, soa**p**, ha**pp**y
ôr	**or**der, f**or**k, h**or**se, st**or**y, p**our**	r	**r**ide, pa**r**ent, wea**r**, mo**r**e, ma**rr**y
oi	**oi**l, t**oy**	s	**s**it, a**s**ide, pet**s**, **c**ent, pa**ss**
ou	**ou**t, n**ow**	sh	**sh**oe, wa**sh**er, fi**sh** mi**ssi**on, na**ti**on
u	**u**p, m**u**d, l**o**ve, d**ou**ble	t	**t**ag, pre**t**end, fa**t**, bu**tt**on, dress**ed**
ū	**u**se, m**u**le, c**ue**, f**eu**d, f**ew**	th	**th**in, pan**th**er, bo**th**
ü	r**u**le, tr**ue**, f**oo**d	t͟h	**th**is, mo**th**er, smoo**th**
u̇	p**u**t, w**oo**d, sh**ou**ld	v	**v**ery, fa**v**or, wa**v**e
ûr	b**ur**n, h**urr**y, t**er**m, b**ir**d, w**or**d,c**our**age	w	**w**et, **w**eather, re**w**ard
ə	**a**bout, tak**e**n, penc**i**l, lem**o**n, circ**u**s	y	**y**es, on**i**on
b	**b**at, a**b**ove, jo**b**	z	**z**oo, la**z**y, ja**zz**, ro**s**e, dog**s**, hous**es**
ch	**ch**in, su**ch**, ma**tch**	zh	vi**s**ion, trea**s**ure, sei**z**ure

actual mechanical advantage (ak′ chü əl mi kan′ i kəl ad van′ tij) of a machine is the ratio between the resistance force and the effort force. Actual mechanical advantage is always less than the mechanical advantage.

effort arm (ef′ ərt ärm) the distance from the fulcrum to the effort force on a lever.

effort force (ef′ ərt fôrs) the force applied to a lever.

first-class lever (fûrst′ klas′ lev′ ər) a lever in which the fulcrum is located between the effort and resistance forces; the direction of effort and resistance is opposite.

friction (frik′ shən) a force that opposes motion.

fulcrum (fül′ krəm) the point on which a lever rotates or pivots.

gravity (grav′ i tē) the mutual force of attraction that exists between all objects in the universe; force exerted by Earth on all objects on or near it.

inclined plane (in klīnd′ plān) a slanted surface used to raise or lower objects; a simple machine.

joule (jül) a unit of work or energy; a newton–meter.

kinetic energy (ki net′ ik en′ ər jē) energy of motion.

lever (lev′ ər) a bar free to rotate around a point; a simple machine; consisting of an effort arm, fulcrum, and resistance arm.

mechanical advantage (mi kan′ i kəl ad van′ tij) the amount by which the applied force is multiplied by a machine.

momentum (mō men′ təm) the mass of an object multiplied by its velocity.

potential energy (pə ten′ shəl en′ ər jē) energy due to position or condition.

power (pou′ ər) amount of work done per unit of time.

resistance arm (ri zis′ təns ärm) the distance from the fulcrum to the resistance force on a lever.

resistance force (ri zis′ təns fôrs) the force resisting the effort force on a lever.

screw (skrü) an inclined plane wound around a cylinder; a simple machine.

second-class lever (sek′ ənd klas lev′ ər) a lever in which the resistance force is located between the fulcrum and the effort force.

simple machines (sim′ pəl mə shēn z) devices that make work more convenient by changing the speed, direction, or amount of force.

third-class lever (thûrd klas lev′ ər) a lever in which the effort force is located between the fulcrum and the resistance force.

watt (wot) the SI unit of power; one watt is one joule of work per second.

wedge (wej) an inclined plane with either one or two sloping sides; a simple machine.

work (wûrk) the transfer of energy as the result of motion of objects; a force applied over a distance.

INDEX

INDEX continued

CREDITS

Photo Credits:

Cover, ©1991 Comstock, Inc.; **1,**©Todd Powell/Profiles West; **3,** (t) ©K.S. Studios, (b) ©Duomo/David Madison 1985; **4,** Brent Turner/BLT Productions; **5,** Mike Powell/Allsport UST; **6,**(b) ©Studiohio 1991, (t) Richard Haynes/RM International; **7,** ©K.S. Studios; **8,** ©Studiohio 1991; **8-9,** M. Angelo/Westlight; **9,** ©Studiohio 1991; **10,** Mitchell B. Reibel/Sportschrome East/West; **11,** Mike Powell/Allsport USA; **12-13,** ©Studiohio 1991; **14,** Stephen & Michele Vaughan; **14-15,** Cary Wolinsky/Stock Boston; **15,** Historical Pictures Service Inc.; **17,** ©Larry Ulrich; **18,** Manfred Gottschalk/Westlight; **21,** Brent Turner/BLT Productions; **22,** Julian Baum/Science Photo Library/Photo Researchers, Inc.; **25,** ©Todd Powell/Profiles West; **26-27,** ©Studiohio 1991; **28-29,** Photo Researchers, Inc.; **29,** Frank S. Balthis; **30-32,** ©Studiohio 1991; **32,** K.S. Studios; **33,** ©Studiohio 1991; **36,** (1) ©Del Mullcey/Photo Researchers, Inc., (r) ©Lee White/Westlight; **36-37,** Brent Turner/BLT Productions; **38-39,** Richard Haynes/RM International Photography; **40,** ©P. DeCesaro; **41,** ©Doug Martin; **42,** Brent Turner/BLT Productions; **44,** (r) ©K.S. Studios, (1) ©P. DeCesaro; **45,** ©Michael Neveux/Westlight; **46,** (bl) ©Doug Martin; (t) ©C.J. allen/Stock Boston; **47,** ©Ron Kimball; **48,** (t) ©Brent Turner/BLT Productions, (m) Chris Springmann/©The Stock Market, (b) Animals Animals/©Stephen Dalton; **49,** Richard Haynes/RM Internatinal Photography; **50-51,** ©K.S. Studios; **51,** (l) ©Duomo/David Madison 1985, (r) ©Platinum Studios; **52-55,** ©K.S. Studios; **57,** (tr)©Doug Martin, (ml) Pete Ceren Photography, (mr) Andy Anderson/Adventure Photo; **58,** ©Platinum Studios; **59,** Andy Caulfield/The Image Bank; **60,** Mark E. Gibson; **62,** ©Platinum Studios; **63,** (t) ©E.R. Degginger, (b) ©P. DeCesaro; **65,** [*]©KS Studios/1992/block & tackle courtesy of The Ohio State University, Physics Dept., Columbus, Ohio; **66,** (t) ©P. DeCesaro, (b) ©Cliff Feulner/The Image Bank; **67,** (t) ©Gary Braasch/Woodfin Camp, (b) ©Graham French/Masterfile; **68,** ©Studiohio 1991; **71,** ©Eric Reynolds/Adventure Photo, (i) Charles Campbell/Westlight; **73,** ©P. DeCesaro; **75,** David Young-Wolff/Photo Edit; **76,** (t) ©Brent Turner/BLT Productions, (b) J.P.H. Images/The Image Bank; **77,** (t) ©David W. Hamilton/The Image Back, (b) Larry Lee/Westlight; **78,** (b) G.V. Faint/The Image Bank, (t) Magnum Photos; **79,** Brent Turner/BLT Production; **80,** Johnny Stock Shooter/International Stock Photo; **80-81,** ©Brent Turner/BLT Productions; **82,** ©K.S. Studios; **83,** David Young-Wolff/Photo Edit.

Illustration Credits:

15, 16, 58, 74, Bill Singleton; **David Reed; 22, 23, 34, 43, 63,** Charlie Tomas; **28, 40,** Stephanie Pershing; **54, 55, 64,** James Shough; **59,** Allan Eitzen; **81,** Henry Hill; **84, 85,** David Bowers